SHINING

in THE

ANCIENT SEA

SHINING *in* THE ANCIENT SEA

❖

THE ASTRONOMICAL ANCESTRY *of* HOMER'S ODYSSEY

LAURIN R. JOHNSON

The author can be reached at
Multnomah House Press
Post Office Box 19687
Portland, Oregon 97280-0687
or, by e-mail at: laurinj@Teleport.com

Front cover photograph by Laurin R. Johnson
Photograph of the author by Michael Ellwein
Cover and book design by Susan Bloom

Library of Congress Catalog Card Number 99-60064
First United States Edition
International Standard Book Number 0-9669828-0-0

Printed in the United States of America
by Ash Creek Press, Portland, Oregon
Printed on recycled paper using soy-based inks.

To my wife, Deborah Rand

With gratitude to
Susan Bloom and Mac Kieffer
at Ash Creek Press
for their encouragement
and advice.

CONTENTS

CHAPTER FIVE ✧ *Okeanos and other Celestial Rivers*

Okeanos is the Milky Way, not a circular river surrounding a flat, disc-shaped earth. The River of Flaming Fire and the River of Lamentation make up the constellation Eridanus.

CHAPTER SIX ✪ *The Northern Path*

The Sirens, Scylla, Charybdis, Planctae, Thrinacie, and Calypso can all be identified with asterisms that lie between the vernal and autumn equinoxes.

CHAPTER SEVEN ✣ *Olive Trees at the Corners of the Year*

The curious double-olive on Scherie and the cave of the nymphs on Ithaca marked the autumn equinox. The olive tree of Odysseus's bed marked the winter solstice.

CHAPTER EIGHT ❖ *Who was Odysseus?*

Odysseus may have begun his long career as the planet Saturn.

APPENDIX - *The celestial location of Ortygia is in the vicinity of the summer solstice.*

NOTES

WORKS CITED

LIST OF ILLUSTRATIONS

LIST OF TABLES

Greek poetry. Greece was waking from a long night of sleep, a night lasting over three hundred years, through which the river of song continued to run like a marvelous dream. In the dawning of a new age the Homeric poems would no doubt have vanished forever had it not been for the advent of writing.

And had it not been for the generations of singers that lived before the introduction of writing, the very memory of an entire civilization, the Mycenaean civilization described in the Homeric poems, would have disappeared. Some four or five hundred years elapsed between the Trojan War, which the Iliad and the Odyssey celebrate, and the appearance of these poems in written form. The river of song flowed uninterrupted through these centuries, and for most of that time through a cultural landscape that can only be described as a wasteland. Shortly after the Trojan War the citadels and palace-complexes of the Mycenaean civilization–Iolkos, Crisa, Gla, Pylos, Mycenae, Tiryns—were destroyed. What followed scholars now call the Greek Dark Age.[1] Archaeology has failed to unearth anything stemming from this period to suggest a level of culture higher than the bleakest subsistence. There is no large-scale architecture. There are no treasures, no luxury articles, no amber, not even many weapons, all of which were abundant before 1200 B.C. Instead, the evidence shows a severe decline in population, general and widespread poverty, and a marked absence of the former powerful and wealthy ruling class. The arts declined, and the art of writing vanished altogether, to be re-introduced hundreds of years later with a different alphabet borrowed from the Phoenicians. The loss of literacy alone is testimony to the cultural devastation experienced by the ancient Greeks, a devastation so complete that not even the memory of it survived. Not a single word about this destruction has ever been found in the legends and myths and ancient writings

INTRODUCTION

The Iliad and the Odyssey are the culmination of an oral tradition that was centuries old by the time Homer was born. A succession of poets who could neither read nor write contributed to an elaborate system by which the Homeric epics were composed, committed to memory, sung, and transmitted from one generation to another over centuries without the use of parchment or writing stylus, but only the voice and ear and rhythmic strumming of a lyre to quicken the memory. As far as scholars have been able to determine, the poems were probably given their final form by an *aoidos*, or singer, named Homer sometime in the ninth, eighth, or seventh century B.C., and were probably written down sometime in the seventh or sixth century B.C. Details more exact than this will probably never come to light, but it is clear by now that the introduction of writing marked not only the beginning of European literature but also the end of a long poetic tradition, a river of song whose headwaters reached far, far back beyond the horizon of history.

The arrival of the Phoenician alphabet in the eighth century B.C. was only one element in a complex of forces that brought about gradual but profound changes in the Greek culture of that time. A dramatic increase in population sent Greeks westward to colonize Sicily and southern Italy, and eastward to colonize Asia Minor and the Black Sea. The polis, or city-state, was beginning to take shape, temples were built for the first time on Greek soil, the Olympic games began, a personal element began to appear for the first time in

HELIACAL RISING

Just before dawn
Turns heaven shallow
The hour between
One dream and the next
When eastern stars
Hang briefly

In the deep sky
Ulysses returns
Still riding the mast
Adrift and shining
In the ancient sea

Charybdis gleams
Above him
Calypso waits
Like time
Below the edge

The hour is perilous

The winged words
Flee like owls
Back to the trees
The dark wine
Fades from the sea

And Ulysses
Old solitary wanderer
Is lost again
In thin, thin air

of Greece, and the Greeks of the classical age were completely ignorant of the catastrophe.

In the past one hundred years archaeology has pieced together the evidence of a civilization destroyed, but how this could have happened still remains one of the great unsolved mysteries of history. At first scholars blamed the collapse of the Mycenaean civilization on the Dorians, a Greek-speaking people who migrated from the north about this time. This view has since been discounted, but no alternative has won universal acceptance yet. One plausible theory, though much disputed,[2] is that Greece suffered a severe and long-lasting drought. Except for those areas on the western slopes of high mountains which continued to receive a trace of moisture, the country simply dried up, crops failed, starvation set in. In this scenario, the palaces were destroyed by the Mycenaean aristocracy's own subjects, driven to desperation by famine. In addition to being the seats of power, the palaces were store-houses of grain and other provender, and the hungry mobs outside the fortress walls were able to overwhelm their well-fed rulers. The neighboring Hittite Empire in Anatolia disintegrated at the same time the Mycenaean civilization collapsed, and a severe drought would seem to be the only plausible explanation for such a wide-spread catastrophe. About this same time Egyptian hieroglyphics record an invasion of the "Sea Peoples," most likely a contingent of Mycenaean Greeks who may have set out across the Mediterranean in search of food and a more hospitable climate. The Egyptians were successful in repelling the invasion, and some of these Sea Peoples, the Peleset, retreated to Palestine where they settled, giving their name to the territory they seized.[3] The Old Testament calls them Philistines, the best known of whom was Goliath, killed by young David with his sling-shot.

The Homeric poems survived the internal warfare, the dispersal of the Mycenaean peoples, the years of bleak subsistence, and finally the return of the rains and with them new prosperity, the dawning of a new culture, and the transition from an oral culture to a literate one. But in the case of the Odyssey, during the course of its long career through the vicissitudes of cultural disintegration, the centuries of suffering and deprivation, and its literary rebirth into a new culture, something was lost. And what was lost was a context in which the poem made sense. What was lost was the answer to a very simple question: *What is the Odyssey about?* The quest for an answer to this simple question is the subject of this book, and the quest will take us back in time before the advent of writing, before the Mycenaean civilization began, before, even, the Greek-speaking people moved down into Greece.

Portland Oregon
April 1999

CHAPTER ONE

✦ *The Puzzle*

A man has been abroad for many years.
Poseidon is always on the watch for him.
He is all alone.
The situation at home is that the suitors
are wasting his money and plotting
against his son.
After a stormy passage he returns,
reveals himself, attacks his enemies,
kills them, and is saved.

(Aristotle, *Poetics*)

Aristotle's very brief summary of the Odyssey is faithful to the theme of homecoming that structures the poem. The "stormy passage," however, comprising books nine through twelve, has remained a puzzle to readers since the days of the Alexandrian commentators in the second century B.C. What are these strange encounters with the Cyclops, Circe, Calypso, the Sirens, and the rest really about? Why this sequence of twelve adventures, seemingly unrelated to one another except in their insistence on the fantastic, when a protracted stay at any one of the strange locales Odysseus comes upon would have sufficed for a tale about a delayed homecoming? Could there be a reason why the adventures happen in the order they do? Did these adventures once contain a meaning now lost to us?

In its career through two millennia the Odyssey has left in its wake countless attempts to find answers to these questions. Heraclitus, a commentator on Homer writing sometime around the first century A.D., called the Odyssey "a vast allegory," and it has been read as such up to and including our own age. The encounter with the Cyclops Polyphemus, for example, was interpreted by one early commentator as wisdom conquering the secrets of nature; another imagined the eye of the Cyclops to be the giant's daughter who burned with love for Odysseus and helped him to escape. Circe's four maids were believed to be the four seasons, Circe representing the year. The wax that prevented Odysseus's men from hearing the song of the Sirens was interpreted as symbolizing the lessons of a wise master which protect the men from their own curiosity. The lengthy imprisonment of Odysseus on Calypso's island symbolized the imprisonment of the soul within the body. And so forth.[1] These allegorical interpretations were the earliest responses to the persistent suspicion that there is more to the Odyssey than a simple tale of one man's return from the Trojan War, a hunch that something else lies beneath the surface. Although they may have satisfied readers of the Odyssey for a time, none of them proved lasting.

In the nineteenth century scholars found a new key to unlock the door to the mystery of the Odyssey. The door came open, but in the dim light of what could be called the first modern exploration, these researchers took the wrong corridor once they got inside. The key was the discovery that most European languages, as well as the languages of India and Persia, were cousins, and the ancestry of all of these languages could be traced back to a single pre-historic language, now called

Proto-Indo-European, and to a single people, called Aryans or Indo-Europeans. Over the past hundred years a number of theories have been proposed about the earliest home of this language, and about the nature of the tribes who spoke it and eventually spread it over Europe and parts of Asia. For a long time the most common belief was that their original homeland was the Caucasus, a region between the Black Sea and the Caspian Sea; hence, those of us most susceptible to sunburn now list our race as Caucasian. The current consensus is that the Indo-Europeans began their career of expansion and conquest around 4000 B.C. from a homeland in the Volga-Ural region to the north of the Caucasus.[2]

As these peoples spread out over Europe and parts of Asia over the next three thousand years the original language took many forms. This tribe or tribes split into western and eastern groups very early, and from the western group eventually derived the Greek, Celtic, Italic, and Germanic languages; from the eastern group stemmed, in time, the Balto-Slavic tribes and the early Persians and Indo-Aryans. The first wave of Indo-Europeans to enter Greece, sometime around 1800 B.C., were already speaking an early form of Greek. The Indo-Aryans who invaded India around 1450 B.C.[3] carried with them the oldest form of the original language, Sanskrit.

Sanskrit is the language of the Vedas, the sacred books of the Hindus. In the second half of the nineteenth century the scholar responsible for editing the English translations of the Vedas was Friedrich Max Müller. In 1856 Max Müller published a small work entitled *Comparative Mythology,* which contained the thesis he was to spend the rest of his life expanding and defending. It was this thesis that sent Max Müller and a host of his disciples down the wrong corridor, and their trail through the labyrinth of mythology, begun with great enthusiasm and high hopes, ended a half century later in almost total disrepute.

Max Müller's fellow philologists had shown that Zeus in Greek, Deus in Latin, Dievas in Baltic, Tyr in Norse, and Dyaus in Sanskrit could all be traced back to the Proto-Indo-European **deiwas*, the name of the sky-god, which in turn came from the root *deiw*, meaning "to shine." This root also lay behind "sky" and "day," and for Max Müller the name suggested the most exalted notion of God possible to the human mind. But given such a lofty notion at the beginning, how could the grotesque character of many Greek myths be explained—Cronus lopping off the genitals of his father Uranus, for example, or Cronus eating his children, swallowing a stone, and vomiting up the whole lot? Max Müller found the explanation in language itself, in the way language, by its very nature, grows and develops. He recognized that the only way to express any new perception is through metaphor, as the original name of God would indicate, or for that matter contemporary expressions like "greenhouse effect" or "global village." At an early period in the development of the language, which Max Müller called the mythopoeic age,

> words were heavy and unwieldy. They said more than they ought to say, and hence, the strangeness of the mythological language, which we can only understand by the natural growth of speech. Where we speak of the sun following the dawn, the ancient poets could only speak and think of the sun loving and embracing the dawn. What is with us a sunset, was to them the Sun growing old, decaying,or dying.[4]

Max Müller saw in this what he called the "disease of language." Metaphor was the only recourse for turning a new perception into a conception, yet the very nature of metaphor invited misunderstanding. These primitive descriptions of the world of nature were understood by the contemporaries of these philosopher/poets, but as generation succeeded generation the original meaning was bound to get lost. What began in the mythopoeic age as a description of the dawn, for example, eventually became in duller minds simply a story about a young maiden, the metaphor misunderstood and forgotten, and with it any mention of the dawn that gave the tale its meaning.

Since myth had evolved along with language and evolved by the laws that govern the growth of language, it seemed possible to trace a myth back to its Indo-European roots the same way one traces the history of a word. It is a path fraught with danger. This early explorer strode confidently into the thicket, but because he believed that the early Indo-Europeans were capable of only the most obvious perceptions and the most primitive kind of reasoning, he found only a small cluster of natural phenomena at the center: sun, moon, day, night, dawn, sunset. This handful of celestial events could be found hidden beneath just about any myth Max Müller turned his attention to:

> While reading the story of Hephaestos splitting open with his axe the head of Zeus, and Athene springing from it, fully armed, we perceive behind this savage imagery, Zeus as the bright Sky, his forehead as the East, Hephaestos as the young, not yet risen Sun, and Athene as the Dawn, the daughter of the Sky, stepping forth from the fountain-head of light.[5]

His theory, supported by a gracious and lucid prose style, attracted a wide audience. The theory held a peculiar fascination, partly because it claimed that by tracing words and myths back to their roots we could watch, in reverse, the development of concepts and of the human mind itself, finally arriving at its birth a few generations this side of Eden. Max Müller became the most famous scholar in Europe, so well regarded that Queen Victoria invited him to Buckingham Palace to instruct her in his ideas, and Oxford University, where he was Taylorian Professor of Modern Languages, freed him of all teaching duties so that he would stay in England rather than accept a post at the University of Strasbourg. His ideas were so popular that by the end of the century Andrew Lang, his most outspoken critic, was to complain that Max Müller's notions were showing up in children's schoolbooks.

One of these notions was the concept of a Solar Hero. Although the Solar Hero had existed in embryonic form in the works of Dupuis seventy-five years earlier, he now emerged fully into the world, and soon he could be found everywhere. A disciple of Max Müller, George William Cox, went on to explain myth after myth in solar terms. Other scholars followed suit, and soon Hercules, Jason, and Odysseus were solar heroes. Sir Gawain of the Round Table, whose fighting strength supposedly waxed till noon and waned as evening approached, became a solar hero. And Siegfried, and Orpheus, and Gilgamesh. Any hero who wandered far and wide or wielded a bright sword became a likely candidate.

The Odyssey, to return to our subject, was taken apart and put back together again by a succession of scholars who were now able to see in the epic tale a primeval story about the Sun. Especially among German scholars, variations were wrought on a simple pattern: Odysseus was the sun, who descends into the Underworld in the winter months and returns in the Spring. Penelope is the Earth-goddess, the 108 suitors are the winter days put to an end by the returning sun. Odysseus's voyage through a sea of troubles represents the sun battling with clouds and with powers of darkness and night. One of the last advocates of this variety of interpretation, J. Menrad, gave this summary:

> The sun god with his twelve companions leaves his palace in the heavens and comes to earth in order to win the great mother-earth as his bride. On his journey he encounters various things that try to thwart

him–the Sirens are a pair of stars that try to trap him with their magic songs, the moon-goddess Circe seeks to win him, the goddess of night Calypso wants to keep him forever in her cave. He overcomes all of the hindrances, since all of his longing is for the earth. He meets Polyphemus, who tries to contain him in his cave of ice, but the Sun-god is stronger, and with his beams melts the crystal eyes of ice and takes away the sheep and goats, the winter storms. But because of this he has made an enemy of the great god of the sea, who tries to prevent him from returning to earth, and a battle ensues between the god of fire and the god of water. With the help of Leucothia (a cloud) he arrives on land and comes as a beggar to the palace. The earth-goddess proclaims that whoever can fire an arrow through the twelve axes, the twelve months of the year, will be her spouse. Odysseus the sun accomplishes the feat, and frees the earth from the tyranny of winter.[6]

We find this kind of explanation of the Odyssey ridiculous today; in fact, it had become ridiculous by the time the nineteenth century came to an end. When Max Müller died in 1900 his theories were being parodied—the tale of four and twenty blackbirds baked in a pie was shown to be a solar myth in disguise, and finally someone was able to prove by Max Müller's methods that Max Müller himself was the Sun.[7] About this time another celestial theory which had enjoyed a considerable vogue in Germany was also collapsing—that of Alfred Jeremias, who believed that the astronomical lore of the ancient cultures of Mesopotamia was the basis for early Near Eastern religions. In addition, he claimed that this lore had spread around the world, and could be found from China to South America. The theory has since been named the Pan-Babylonian Heresy. With the collapse of these two grand theories and the ensuing ridicule, astronomical elements in myths, and particularly in Homer, became anathema. A host of learned scholars had been deluded, and the succeeding scholars have stayed gun-shy for nearly a century. They have consistently ignored the stars, as have most of us in this age of electricity. Once in a great while one or another writer will hesitantly bring up the subject, as in this 1962 footnote: "For a long while so much misuse was made of alleged sun and star myths that this species of explanation has been suspected and condemned. It is time the ban were broken."[8] Indeed, it may turn out that, although the conclusions of these scholars were misguided, the hunch that we should look to the heavens for an answer to the mystery of the Odyssey was not so ridiculous after all.

Rather than looking to the heavens, one recent scholar in particular has looked within the human psyche for a solution. Joseph Campbell has found in the Odyssey an allegory of the Jungian process of "integration," in which the masculine psyche of a patriarchal culture comes to terms with the subconscious feminine nature lying dormant, "submerged like an Atlantis." Odysseus's voyage is the Night-sea Journey, with Hermes, the guide of souls to the underworld, as the patron god. For Campbell, "Odysseus's meetings with Circe, Calypso, and Nausicaa represent psychological adventures in the mythic realm of the archetypes of the soul, where the male must experience the import of the female before he can meet her perfectly in life." In this reading of the Odyssey each adventure is interpreted as a necessary step on the path to self-knowledge. The sojourn in the land of the Lotus-eaters, for example, represents forgetfulness of the everyday and the turning of the mind to the inward realm; the Cyclops is the cosmic threshold guardian, his eye analogous to the "sun door" leading to the "sphere of transpersonal forces over which the ego has no control;" Circe is the guardian of the Underworld, where Odysseus meets Tiresias, who has himself been both male and female and thus possesses the secret of immortality.[9]

Admittedly, the Odyssey does seem to have a 'feminine' quality about it, especially in contrast to the Iliad. Samuel Butler even maintained that the Odyssey was written by a woman.[10] It may well be true that the Odyssey betrays an archetypal message, one that even its authors were unaware of, an allegory of the male psyche coming to terms with its suppressed feminine half. Yet there are other reasons why Odysseus encounters such a preponderance of females on his journey, reasons having nothing to do with the struggles of the human psyche with itself. We shall get to them in due time.

Many commentators who have little patience with allegory prefer to take the Odyssey at its word, and see a description of an actual voyage in Books 9-12. To these commentators the cave of the Cyclops or the island of Calypso or the whirlpool of Charybdis are actual places, and one ought to be able to follow the trail of Odysseus, sail where he sailed and walk where he walked. Like the search for hidden allegorical meanings, the search for these actual places has been going on since ancient times. Herodotus, Thucydides, Apollonius, and Virgil all make reference to various locations around the Mediterranean and their supposed connection with the Odyssey, the most complete list being given by the geographer Strabo. In our own times the search has been renewed, notably by the French scholar Victor Bérard[11] and the English ex-naval officer Ernle Bradford,[12] who both follow Strabo in searching out locations in the western Mediterranean. While these two scholars agree on the location of the Lotus-eaters (Jerba), the Laestrygonians (Bonifacio), Scylla and Charybdis (the Straits of Messina), and Circe's palace (Circeo), they are scarcely in accord about many other particulars. Other possible locations around the Mediterranean have been found by Mauricio Obregon,[13] while Gilbert Pillot[14] argues that the Odyssey is an account of travels to the northern Irish and Scottish coasts, and A. Rousseau-Liessens[15] argues that Odysseus was really sailing the Adriatic. The most recent attempt at tracing the itinerary of Odysseus has been made by Tim Severin, who sailed in a 54-foot replica of a late Bronze Age galley to prove that Odysseus was seldom out of sight of the Greek coastline.[16]

By the latest count, some seventy theories about the geography of the Odyssey have been proposed.[17] These attempts to establish an itinerary for Odysseus's wanderings rely on Homer's descriptions of the lay of the land and the winds that blow him from one place to another. But these descriptions are so meager when they are given at all that ample room is left for conjecture and dispute, and in the end one guess is about as good as another. Moreover, these theories all overlook what is most fascinating and memorable about the wanderings of Odysseus: the strange, grotesque, horrible, sometimes enchanting creatures he encounters. It should be obvious that creatures like the Cyclops, the giant Laestrygonians, the six-headed Scylla, the goddesses Circe and Calypso, or Aeolus, steward of the winds, are not to be found on any island washed by the waters of any earthly sea; they are creatures from another dimension, from the vast sea of folktale and myth that lapped its dark waters around the edges of every ancient mind.

To view the Odyssey as a collection of folktales rather than as an unguided tour of the Mediterranean certainly makes more sense. In recent years many scholars have turned their attention to isolating the elements of folktale in Homer's epic and to searching out motifs and parallel tales in the folklore of other lands. The results of this search suggest that the adventures of Odysseus are not original with Homer, and perhaps not even original with the Mycenaean culture Homer sang about. Over two hundred versions of the tale of the Cyclops have been found and few of these, if any, could be said to have derived from Homer's Odyssey.[18] The tale of the Lotus-eaters, who offer a food that, when eaten, prevents the traveller from returning home, finds its counterparts throughout the world. The Sirens, Circe, and the Wandering Rocks or Symplegades which Circe describes to Odysseus are all found in the saga of Jason and the Argonauts, a pre-Homeric tale of considerable antiquity. Circe's house, standing in a clearing in a forest dell and sighted from a distance by its column of smoke, belongs more to the forests and folklore of Central Europe than it does to the shores of the Mediterranean.[19] One scholar has suggested that the story of Odysseus's bed, built with a living tree for a bedpost, may go back to neolithic times.[20]

As our investigation unfolds it should become evident that most of the tales in the epic had their birth in that long night before the dawn of history. For now, we can at least conclude that Homer did not invent them, and any attempt to interpret the Odyssey as an actual voyage around the Mediterranean at the close of the second millennium B.C. is akin to setting out on an expedition in search of Noah's Ark, or investing in a metal detector to look for the Holy Grail.

Suppose we agree with a number of contemporary scholars that the central section of the Odyssey, the wanderings of Odysseus, is nothing more than a collection of separate folktales passed from memory to memory through generations of story-tellers and singers and finally woven together around the character of Odysseus by Homer or one of his predecessors. Implicit in this explanation is the belief that there is no significance in the number of tales collected. One tale more or one tale less would hardly matter in the long sweep of the poem. This explanation also implies that there is no necessary order to the tales. There need be no reason why the encounter with Aeolus should come after the tale of the Cyclops rather than before, or why the encounter with the Laestrygonians should come between the story of Aeolus and the story of Circe. If these adventures are merely a haphazard collection of yarns, we would not expect to find a significant pattern in their number (twelve) nor in their sequential order. But their number and order do make up a pattern, a design that is astonishing in its symmetry and complexity. [figure 1.1]

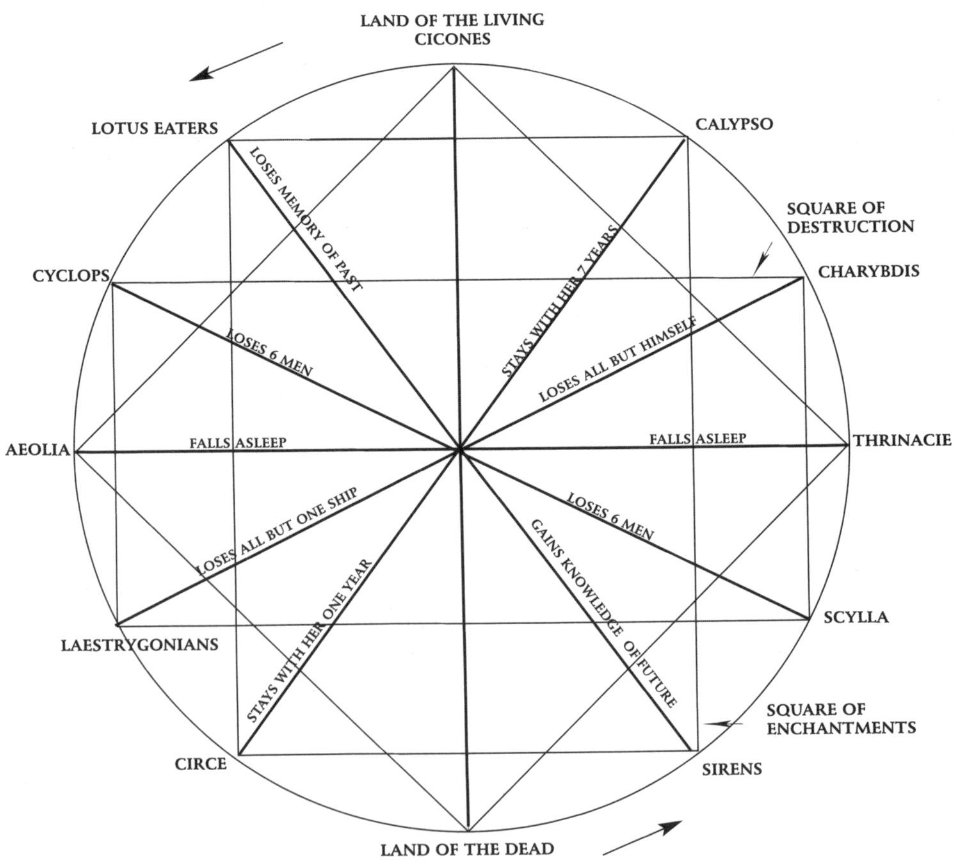

Figure 1.1 Geometric Structure of the Odyssey

The geometric structure of Books 9 through 12 of the Odyssey is indeed curious. Each adventure on the circle is mirrored in some way by the adventure opposite: Odysseus falls asleep after leaving the island of Aeolia, and while he sleeps his men open the ox-skin-bags containing the winds, preventing him from reaching Ithaca; on the opposite side of the wheel Odysseus again falls asleep on the Island of the Sun, and his men slaughter the oxen of Hyperion, causing their own death and, as Tiresias warned, delaying his return. Odysseus sleeps with two goddesses, Circe and Calypso, the one at the outermost edge of the world, the Island of the Rising Sun, the other at Ogygia, the "Navel of Ocean;" Circe's palace is the entrance to the dark Land of the Dead, Calypso's cave the way back to the familiar daylight world of the living. The Cyclops devours six of Odysseus's men, Scylla devours another six, and so on around the wheel.

Within this wheel of opposites and correspondences are other patterns: the square of enchantment, formed by the Lotus-eaters, Circe, the Sirens, and Calypso, in which escape from the wheel of Time is promised and resisted; and the square of destruction, formed by the monsters of the saga, the Cyclops, the Laestrygonians, Scylla, and Charybdis, whose gruesome dismembering, devouring, and vomiting all present death and time at their most ravenous. These two squares are embraced by a third, formed of the primal powers—the sun at Thrinacie, the winds at Aeolia, and at the Land of the Dead the deep-flowing River of Ocean that circles the world.

This circular structure of Odysseus's account of his wanderings is repeated in the narrative structure of the poem. Homer begins the poem where it will end, in Ithaca. The great circle of Odysseus's departure for Troy, his absence of twenty years, and his final return frames another, the departure and return of his son

Telemachus; and the cycle of Telemachus's journey contains within it accounts of other homecomings, those of the Atreides, Agamemnon and Menelaus, and serves to frame, in its turn, Odysseus's own account of his wanderings. This account forms the innermost circle of the poem, beginning with the description of his stay on Calypso's isle, and ending the full account of his travels again with Calypso, who serves as both introduction and conclusion to his tale.

Parallel to the geometric shape given to the plot of the Odyssey is the circle of Odysseus's fortunes, beginning in riches and power, twelve ships and men to sail them, losing, as the wheel turns, all but one ship, and then that, too, finally washing up on the beach at Phaeacia with nothing but his life, and barely that, naked and nearly dead, covering himself with leaves to keep the spark of life alive in him. At the nadir of his fortunes he is brought back into life with the help of the maiden Nausicaa, and regains, step by step, his wealth, his herds, his family, and his kingship.

These circular frames around the voyage of Odysseus are loosely constructed. What is surprising is the tight geometric structure of the voyage itself. How are we to account for it? Why is the geometry there at all?

One answer might be that Homer and his predecessors were echoing the same sense of form that distinguished the visual arts of their day, seen most clearly in the style of vase-painting that began about 1000 B.C. and continued until around 700 B.C. After the break-up of the Mycenaean civilization, the curvilinear patterns and representations of plants and marine life that characterized the Mycenaean-Minoan style gave way to formal geometric designs—zigzags, shaded triangles and squares, concentric circles, semicircles, rosettes, wheel ornaments, swastikas, and extremely stylized human and animal figures. This

300-year period in Greek art, coinciding with the accepted date of Homer's songs, has since been named the "Geometric Period."

While it is true that the spirit of any age usually informs the different modes of art, as, for example, the correspondences between the architecture, music, and prose style of eighteenth century Europe, in the case of Homer the more subtle aspects of the geometric style go beyond what any listener would be able to perceive. Certainly the larger circles in the design of the Odyssey, the circle of Telemachus's journey framing the circle of Odysseus's wanderings, for example, might have delighted the aesthetic sensibilities of his age. But the balances and correspondences around the circle of the wanderings shown in the diagram would have been imperceptible to Homer's audience in the ongoing course of his song. It is a visual pattern rather than an aural one, static rather than sequential.

This is not to say that there is no aesthetic intent in the structuring of the innermost wheel. The rhythm of the sequence of adventures, tales of gruesome death-dealing monsters alternating with tales of enchantment and these set off by periods of respite, give a music to the narrative, a music that would have been lost if the tale of the Laestrygonians followed on the heels of the Cyclops, lost if Odysseus had landed in Calypso's arms immediately after leaving Circe. This structuring may indeed have been influenced by the geometric aesthetic, in ways that would be impossible to sort out. But for the oral poet there is a far more compelling reason to arrange the elements of the narrative in a visual pattern, and that reason is mnemonics. In addition to memorization through repetition, aided by a prodigious store of poetic formulae, deliberate memory techniques must have been employed, techniques that enabled the poet to conceive of the poem as a whole so that he could know where he was and what was to come next at any given moment of the song. The technique of memory most likely to have been employed was the visual image, a spatial pattern on which to weave the thread of song, a pattern easily held in the mind's eye.

I am not suggesting that the ancient bards ever imagined anything like the diagram proposed here. But whatever the technique, it is apparent that the sequence of adventures is not arbitrary, and that great ingenuity was employed to retain all of these adventures, and retain them in their proper order. But this simply gives more urgency to our original question: *What is the voyage of Odysseus about?* Why such an elaborate mechanism for remembering a sequence of events that are obviously not historic in nature, not the history of a tribe or its genealogies or stories about the mountains and valleys and sacred places familiar to the Greeks? On the contrary, the events in the Odyssey are characterized by their remoteness, expressly not about the known geography of the Greek world nor, obviously, any kind of accurate accounting of what might lie in the western Mediterranean. What is it that is being remembered here?

The answer lies in another great circle, wheeling over our heads at this very moment—the circle of the ecliptic, the path of the sun and moon and planets among the stars. Beneath the Bronze-age trappings of the Odyssey lies an archaic astronomical text that existed before the Trojan War and the ensuing Dark Age, existed, in fact, before the Indo-Europeans migrated out of the Russian Steppe down into India and Greece, before the Indo-Europeans separated into their Indo-Iranian and European branches. The most striking evidence for this can be found in the correspondence between the sequence of adventures recounted in the Odyssey and the peculiar system of astronomy found in the Vedic literature of India. The correspondence is exact, as will be demonstrated in the following chapters, and can only be explained by a common source for both the Greek and Indian traditions. When the separation into the two major Indo-European branches took place has still not been established conclusively. We can be certain, however, that the astronomical lore that gives the Odyssey its structure was at least five hundred years old by the time the epic was written down, and some patterns that show up in its intricate weaving indicate that archaic astronomers were casting their observations in memorable forms a thousand years before that. The poetic tradition that came to an end with Homer's Odyssey extends so far back in time as to defy our powers of belief or even comprehension. Yet the Odyssey contains unmistakable evidence that it is so.

Rather than the abstract language of mathematics, the ancient Indo-European astronomers recorded their observations in the only terminology available to them—a story with human actors–and transmitted these observations by the only means at their disposal—poetry.[21] The language is nevertheless a technical language, and will appear to our eyes, because so far removed in time, as a kind of code.[22] Most of the images that have puzzled Homeric scholars—the plant called *"mōly"* that Hermes gives to Odysseus, for example, or the Cave of the Nymphs on the shore of Ithaca, a cave with two mouths, one for men and one for the gods—are part of the archaic astronomical terminology. The code, once understood, will reveal an astronomy far more sophisticated than ever imagined.

When we contemplate the phenomenon of a body of knowledge passing from human memory to human memory across several thousand years, and that knowledge evincing a continued practice of astronomical observations, our immediate reaction is to say that it wasn't possible, especially when the very language which was a vehicle for the tradition changed on the tongues of the poets from generation to generation until the last of the poets in the two great tribes could not have told they were speaking varieties of the same language. Proof that it was in fact possible stands at Stonehenge, which was in use in various stages of construction for at least 1500 years, from early in the third millennium until around 1200 B.C. When Gerald Hawkins announced some thirty years ago that Stonehenge was actually an astronomical observatory,[23] the chorus of outrage and disbelief among the scientific community was astounding. It could not possibly have been so, because the people who roamed the Salisbury Plain in the third millennium B.C. were impossibly "primitive." R.J.C. Atkinson, initially the most outspoken of the critics out to disprove Hawkin's discovery, had the courage and humility eventually to admit he was wrong. He was ultimately convinced by the evidence, although he was left with an even bigger problem:

> We are dealing here with a population whose average life-span was very markedly less than our own, to the extent that even the 18.6 year period of rotation of the lunar nodes would not be completed even once in the working life of a trained observer. There must, therefore, have been some means of transmitting stored data from one generation of observers to the next and for this reason alone one might expect it to have been recorded in durable form. That it was not recorded in this way leads us to the residual hypothesis that it was stored and transmitted orally. The difficulty of doing this is sufficiently imaged by supposing that the Astronomical Ephemeris is not published, but is transmitted by word of mouth from its compilers to its users in the form of epic verse which must be strictly memorized and reproduced. I can do no more than state the problem. I have no acceptable solution to offer.[24]

The solution has been there on the bookshelf all along, standing among the greatest monuments to the human spirit—the Odyssey of Homer. Not that a distant ancestor of Homer's Odyssey was chanted by the ancient astronomers who built Stonehenge and maintained it as an observatory for some 1500 years. But if not a proto-Odyssey, something very like it would have had to be memorized and sung and passed on to generation after generation of astronomers standing under the canopy of stars in those distant and silent nights.

A counterpart of the Odyssey is not to be found in other later epics like the Aeneid or Paradise Lost. A closer analogy would be Stonehenge itself, or perhaps a medieval cathedral, Notre Dame at Paris, for example, constructed by generations and generations of men holding to the same conviction, honing a complex architecture to resemble the mysterious heartbeat of the universe. The difference is that instead of the two hundred years it took to build Notre Dame, the cornerstone of the Odyssey was laid several thousand years before the last icon was finally cemented in place.

If we examine the foundation of both Notre Dame and the Odyssey we will discover that they are both laid out with reference to the heavens. They both look East to the horizon where the sun and moon and stars are seen to rise. Thus, before we embark on a detailed analysis of Odysseus's voyage among the stars we need to leave the book on the desk for an evening and step out in the back yard, and acquaint ourselves with the methods used by those ancient astronomers who recorded their observations in the only lasting medium available to them—poetry.

CHAPTER TWO

☆ *The Celestial Framework*

Come all you fair and tender maidens
Take warning how you court young men
They're like the stars on a summer morning
They'll first appear, and then they're gone

Appalachian Folk Song

If we were to step out into the backyard just after sunset and watch carefully, we would witness the light of the day continue to linger close to the surface of the earth for a little while, soft, diffuse, coming from no direction, casting no shadows. We expect the darkness to descend, but it begins in a pool at our feet, coming up from the ground, filling the low places, rising until everything is softly submerged and we are seeing more by memory than by eyesight. The light has retreated to the horizon, where it gathers in a great, pale circle, hardly brighter in the west than in the east. Directly overhead the blue of the sky darkens, and the first stars of night begin to appear, one by one in sudden, faint points of light. Slowly the darkness flows down the walls of heaven, until at last the horizon is subdued and the firmament is visible. Night has fallen. We have passed through the familiar gates of twilight and find ourselves in a different dimension. The sky is now vast and deep and bowl-shaped, whereas an hour before it had been empty and without proportion, shapeless, the only reference point the sun. Now, instead of a single ball of fire moving across the sky, the sky itself seems to move, turning slowly around a motionless center, the pole star.

Stars close to the pole revolve in small circles in the course of a night, while those in the southern sky sweep in great arcs from horizon to horizon. Around midnight those stars which first appeared directly overhead after sunset begin to vanish below the western horizon, while in the east new stars appear and climb slowly towards the zenith. As dawn approaches, the latecomers to the night sky, the stars that are rising just ahead of the sun, appear over the eastern horizon and shine for only a brief moment before the light of the coming day causes them to fade away. This brief glimmer is what our lovely Appalachian folk song refers to—a star first appears, and then it's gone. This is called by astronomers the heliacal rising of a star—heliacal from the Greek *helios*, or sun.

If a star lies close to the path of the sun, it will not have been visible at all for forty days and forty nights before its heliacal rising, having last been seen just after sunset, coming into view low in the west as the sky darkens, appearing only briefly before it followed the sun below the horizon. This final brief appearance in the western sky is called the heliacal setting. The next evening the star will not be seen at all, having set too soon to be visible, but forty days later it will reappear briefly on the eastern horizon just before sunrise, and on subsequent days will rise earlier and earlier, climb higher and higher in the night sky before the sun appears and dispels the darkness.

There is more to the heliacal rising of any given star than the spectacle of a brief pinpoint of light on the eastern horizon. The star's reappearance in the eastern sky marks a specific day on the great circle of the year.

On a particular day every year, just before sunrise, at a particular point on the horizon, our absent star will come back. A year has passed since its last heliacal rising; it will be exactly a year before the same spectacle recurs. By watching the horizon at dawn or at sunset, we can locate ourselves in time. In Egypt the heliacal rising of the Dog Star Sirius was a signal that the annual flooding of the Nile was about to begin. Homer's contemporary, Hesiod, speaks of the times of risings and settings of the Pleiades and the star Arcturus as though they were commonly observed, part of the everyday knowledge of his time:

> At the time when the Pleiades, the daughters of Atlas, are rising,
> begin your harvest, and plow again when they are setting.
>
> The Pleiades are hidden for forty nights and forty days,
> and then, as the turn of the year reaches that point
> they show again, at the time you first sharpen your iron.
>
> *(Works and Days, 383-387)*
>
> Now, when Zeus has brought to completion sixty
> more winter days,
> after the sun has turned in his course, the star Arcturus,
> leaving behind the sacred stream of the ocean,
> first begins to rise and shine at the edges of evening.
>
> After him, the treble-crying swallow, Pandion's daughter,
> comes into the sight of men when spring's just at the beginning.
> Be there before her. Prune your vines.
>
> *(Works and Days, 564-570)*
>
> When Orion and Seirios are come to the middle of the sky,
> and the rosy-fingered Dawn confronts Arcturus, then, Perses,
> cut off all your grapes, and bring them home with you.
>
> *(Works and Days, 609-611)*[1]

The heliacal rising of any star can be used to mark the passing of a year's time. But where in the endlessly circling heavens shall we mark the year's beginning? Any point we choose would serve, theoretically, but the obvious choice is at the season of rebirth—either the rebirth of the light in the dark winter days, now our Christmas celebration, or the rebirth of nature in the Spring, now our Easter. Either of these points can be determined precisely by observing the position of the sun as it rises on the eastern horizon.

On successive mornings the sun will rise at a slightly different point on the horizon. As summer approaches the sunrise moves further and further to the north until, like a pendulum, the movement slows, stops, and then reverses, the sun slowly beginning to move towards the south. The sun will continue to rise further and further south on the horizon until midwinter, when it will turn and begin its slow movement towards the north again. The points at which the sun stops and turns (in Greek *trope*, from which we get our term tropic) are called the solstices (from *sol*—sun, and *stitium*—to stand still). Early in the first millennium B.C. the position of the sun at the solstices coincided with the constellations Capricorn and Cancer, hence the familiar terms Tropic of Capricorn, Tropic of Cancer.

The precise moment of the solstice is difficult to determine using sightings on the horizon because, like the pendulum slowing until it stops and reverses direction, the shift in the point of sunrise around the time of the solstice is nearly imperceptible. Unless one had very precise measuring devices, a flat horizon with long sightings (and, we might add, good weather), one would be hard put to tell which day out of four or five marked the northern-most or southern-most sunrise. Two other moments in the great circle of the solar year can be located more easily and with greater precision than the solstices, because the distance between the rising points along the eastern horizon is much greater on successive mornings. These moments are the equinoxes (from *aequi*—equal, and *nox*—night), the two times during the year, once in the spring, once in autumn, when the day and night are of equal length, the two days in the year when the sun rises due east and sets due west.

If markers, say two fenceposts or two large stones, have been erected on the horizon to fix the northern-most and southern-most points of sunrise, another marker placed exactly half-way between the two would point to the sunrise at the moment of equinox. [figure 2.1]

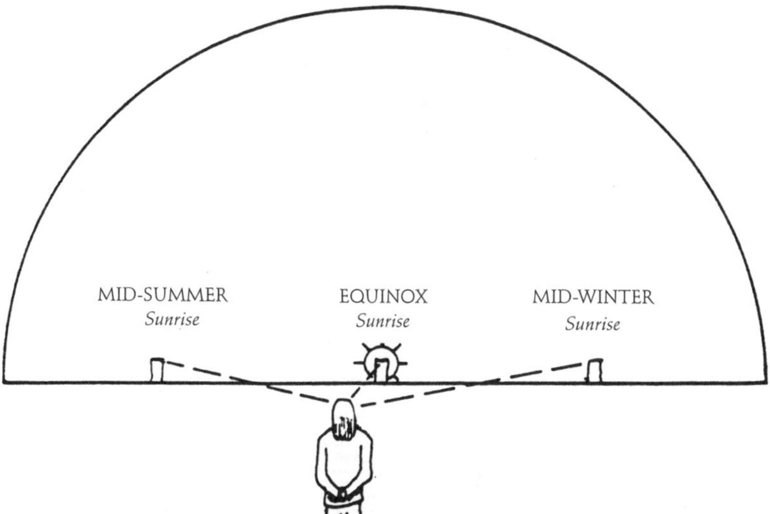

Figure 2.1

Sighting Markers on the Eastern Horizon

After a few years of sightings, a particular star whose heliacal rising would announce the moment of equinox could be determined, and the equinox would be identified with that star. Other stars which would herald the autumn equinox and the winter and summer solstices could also be found by the same careful observation, and these four stars will then mark the four corners of the circling year. We might imagine them as 3-6-9-12 on the celestial clock, the hour hand being the sun, ticking around the twelve zodiacal signs in the course of a year. Early in the third millennium B.C. astronomers in the Tigris-Euphrates valley established these four corners in the constellations Taurus, Leo, Scorpio, and Aquarius, and these four signs persist as "fixed" signs among contemporary astrologers, even though they no longer serve to indicate the solstices and equinoxes. Astronomers in ancient Persia singled out four bright stars, called the Four Royal Stars or the Four Guardians of Heaven, to mark the corners of the year. These were Aldebaran in Taurus, Regulus in Leo, Antares in Scorpio, and Fomalhaut in Pisces Australis.[2] The four corners can be symbolized by a simple cross, and each of the corners can be associated with one of the four directions: Spring–East; Summer–North; Autumn –West; Winter–South. [figure 2.2]

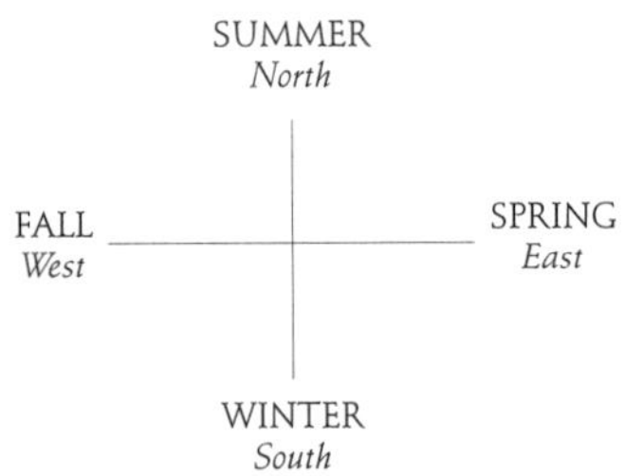

Figure 2.2 The Four Corners of the Heavens

But since the heavens rotate, the slow wheeling of time marked by the solstices and equinoxes is more appropriately symbolized by the swastika. The image of the swastika suggests the frame of a cosmos—the four cardinal points circling around a fixed center. Elements of this very ancient cosmology turn up, in various stages of elaboration and decay, in every culture whose roots can be traced to Indo-European beginnings, whether Celtic, Norse, Prussian, Roman, Greek, or Indic. The oldest written texts containing references to this astronomical knowledge are the sacred books of the Hindus, the Vedas. These books are the culmination of a long oral tradition which seems to have been more rigorous in India than in the rest of the Indo-European world. It is, then, in the Hindu Vedas that we can expect to find a record closest to the original Indo-European conception of the heavens.

The term Veda means "knowledge," and comes from the Indo-European root *weid*, "to see," from which derives the English "wisdom." The Vedas are comprised of four collections (samhitas) of sacred writings: the Rig-Veda, or "Veda of Praise," containing 1028 hymns; the Sama-Veda, a collection of melodies; the Yajur-Veda, a collection of sacrificial formulas divided in two parts—the Taittiriya-Samhita and the Vajasaneyi-Samhita; and the Atharva-Veda, containing magical formulas, myths, and legends. Various later compositions called Brahmanas and Upanishads have been appended to the Vedas.

None of these collections contains a treatise on astronomy as such, although there is ample evidence scattered through the Vedic literature of a knowledge of the stars and of the path of the sun and moon and planets among them. From the scattered evidence it is possible to reconstruct, at least partially, the pre-historic interpretation of the celestial events circling overhead in the night sky, although Sanskrit scholars have sometimes differed among themselves about the meaning of certain terms, especially when the terms occur in passages riddled with metaphors, like the following:

> Twelve fellies, a single wheel, three naves;
> Who can understand it?
> Three hundred pegs have been hammered in,
> And sixty nails which none can move.
>
> *(Atharva Veda 10.8.4)*

The 360 pegs and nails obviously refer to the days of the year, but a single wheel with three hubs and twelve rims has understandably left behind it a trail of conflicting interpretations. The problem is compounded by differing notions about the antiquity of certain Vedic hymns and differing presuppositions about the intelligence and sophistication of the Vedic priests and poets. Those scholars unwilling to grant much of the latter point to passages in the Vedas which claim that the distance from the earth to the sky is equal to 1000 buffaloes standing one on top of the other. They maintain that nothing more than the rankest superstition filled those ancient minds, and what astronomy there is in the Vedas must have come from the second millennium B.C. Other scholars, willing to imagine that the stars had been carefully observed long before then, point to passages that suggest otherwise. For example, it is known that *dhruva*, in the literature of the Vedas, denotes the pole star. In the Sutras this star, the name of which means "firm—fixed—constant," is mentioned in connection with the marriage ritual as an emblem of constancy that the bride should imitate. *Dhruva* does not refer to our Polaris, which has only been close enough to the actual pole to serve as a pole star for the past thousand years or so. In Homer's time the star closest to the pole was Kochab, located on the outer lip of the "Little Dipper," Ursa Minor. But this star never got closer than six degrees from the true pole, and could hardly be said to be a pole at all, since in the course of 24 hours it would have described a circle with a diameter over twelve times the apparent diameter of the moon. For a star to appear at the pole as "fixed" or "constant" we have to turn the clock back to about 2750 B.C. (give or take a few hundred years) and our gaze to the star Thuban (alpha Draconis) which was then less than 10' from the true pole.

In the Maitrayani Upanishad, composed in the first millennium B.C., the movement of *dhruva* is mentioned (5.33.10). Sanskrit scholars have been at a loss to explain this, since to take the passage at face value one would have to acknowledge that astronomical observations had been carried on for two thousand years, and that a record of these observations had been passed from memory to memory through all the generations in that span of time. It is indeed hard to imagine, but any other conclusion seems equally hard to maintain in the light of another feature of Vedic astronomy—the list of asterisms that define the ecliptic, the path of the sun and moon through the stars.

Everyone knows that the moon orbits the earth, the earth orbits the sun. How could the sun and moon be said to travel a path through the stars? This notion may take some explaining. If we were able to look over the sun's shoulder, as it were, from different points along the earth's orbit, that is, at different times of the year, we would see from each point a different constellation of stars behind the sun. At one time the sun would appear to be standing in the constellation Leo, say, and at another time the constellation Scorpio would appear behind the sun. Ancient astronomers could not have imagined such a thing, inhabiting as they did a universe in which the earth stood still and the sun did the orbiting. Neither could they (or we) see both the sun and the stars at once. But by watching the horizon morning and evening, in effect looking over the sun's shoulder, they could see as the year revolved which constellations took turns standing above the sun at sunrise or sunset. All of them together form a great circular path, half of the circle visible at any time of night, the other half below the horizon. It is the same path the moon travels in her monthly rounds, the same path on which the planets can always be found, creeping along at their slower paces.

Different cultures have differently named this path through the stars that make up the great circle of the year. The Akkadians called this path *Pidnu-sha-Shame*, the Furrow of Heaven. Aristotle called it *o Kuklos ton Zodion*, the Circle of Little Animals, from whence our term Zodiac. Aristotle's zodiac originated with the Babylonian astronomers, and is made up of that sequence of constellations familiar to us from reading the astrology column in the daily newspaper: Aries, Taurus, Gemini, Cancer, Leo, Virgo, Libra, Scorpio, Sagittarius, Capricorn, Aquarius, Pisces. The Egyptians, instead of these twelve "houses of the sun," developed a system of decans, that is, a series of 36 stars which rise heliacally at roughly ten-day intervals, completing their cycle in 360+ days. The Vedic astronomers, along with the Chinese, imagined a different face on the celestial clock, and named 27 (later 28) asterisms to mark the progress of the moon across the sky. The moon takes 27 or 28 days (actually 27.3216) to complete its circle of the heavens, moving from one of these asterisms to the next with each passing night. These asterisms were called *naksatras*, a Sanskrit word meaning "having rule over night."[3]

In the Vedas the list of *naksatras* commonly begins with *krttikas*, the Pleiades. The only reasonable explanation for the fact that the Pleiades were chosen as the beginning point of a continuous circle rather than any of the other 27 asterisms must be that the

Pleiades marked the beginning of the year and the yearly round of ceremonies and sacrifices. The Pleiades did once serve this function between 3000 B.C. and 2200 B.C., 3000 B.C. being the earliest and 2200 B.C. the latest possible date that the Pleiades marked the vernal equinox, an era that coincides with the pole star Thuban, or *dhruva.*

There are scattered references in the Vedas[4] indicating that the *naksatra*s were sometimes thought of as being wives of the moon, Candra, a masculine deity in the Vedas. The moon, according to the myth, spent each night with a different wife around the circle. A residue of this notion lingers in the Odyssey, which is remarkable for the number of females, human and otherwise, that Odysseus encounters: Circe, the Sirens, Scylla, Charybdis, Calypso, Nausicaa and her mother Arete, Eurycleia, Penelope, and, of course, Athene. It does not seem far-fetched to conclude that the feminine quality of the Odyssey is more likely to be a consequence of its celestial beginnings rather than its supposed feminine authorship.

Each *naksatra* is named, of course, the names cryptic evidence of a great sea of stories long since vanished, leaving behind only bits and pieces on the shore in fragmentary names like "the bearer-away," "the hide," "dawn," "blind," "moist," "restorers of wealth," etc. We can imagine that each *naksatra* was the focus of a story, or perhaps many stories. Most likely these stories were woven together in various ways. At any rate, the Odyssey, as we shall see, preserves one such weaving.

One of the many puzzles in Vedic mythology is the creation of space by Visnu, who accomplished this by taking three strides. In a well-known hymn of the *Rig Veda* (1.154) we read:

> Let me now sing the heroic deeds of Visnu, who has measured apart the realms of earth, who propped up the upper dwelling-place, striding far as he stepped forth three times.
> They praise for his heroic deeds Visnu who lurks in the mountains, wandering like a ferocious wild beast, in whose three wide strides all creatures dwell.
>
> Let this song of inspiration go forth to Visnu, the wide-striding bull who lives in the mountains, who alone with but three steps measured apart this long, far-reaching dwelling-place.
>
> His three footprints, inexhaustibly full of honey, rejoice in the sacrificial drink. Alone, he supports threefold the earth and sky–all creatures.
>
> Would that I might reach his dear place of refuge, where men who love the gods rejoice. For there one draws close to the wide-striding Visnu; there, in his highest footstep, is the fountain of honey.[5]

The Sanskrit verb in the first verse meaning "to prop up," *skambh*, is related to the noun *skambha,* "pillar," the *axis mundi* propping up heaven. The *skambha* has as its pinnacle the pole star; consequently it is sometimes imagined as slanting towards the north, propping up the sky "as a buttress props up a rampart" (RV 4.5.1). This much is clear. But what is symbolized by the three strides of Visnu has long puzzled scholars. Dawn, noon, and sunset have been proposed, as well as earth, atmosphere, and sky. Yet it is difficult to see how either of these triads could measure anything, or how honey-filled footprints could fit into either interpretation. The three strides do, however, suggest our three markers on the eastern horizon which locate the solstices and equinoxes among the stars. Interpreting the three strides as being these three markers does do justice to the notion of measuring the universe. By them are measured time, the four corners of the year, and space, the four directions.

This conception may sound rather exalted for pre-historic astronomers, and I would hesitate to put forward such an interpretation of the three strides if it weren't for the honey-filled footprints left behind. In the Odyssey honey turns up in the cave already mentioned in Chapter One, the cave with two mouths, one for gods and the other for men. This cave, as I shall demonstrate in due course, is an unmistakable symbol of the asterism that once marked the autumn equinox, and this asterism could only have been located by those prehistoric astronomers by sighting across three markers on the eastern horizon. Honey also shows up in another cosmological story, the Biblical tale of Samson, whose name is a derivative of *shemesh,* "sun." In that story, which I shall also get to in due course, the strange riddle about honey found in the carcass of a lion (Judges 14:5-18) may have derived from an ancient reference to the summer solstice.

Most important for our understanding of the astronomical roots of the Odyssey is the curious "double journey" of the sun mentioned in several of the creation hymns in the Rig Veda:

> The generations have become wise by the powers of him who has propped apart the two world-halves even though they are so vast. He has pushed away the dome of the sky to make it high and wide; he has set the sun on its double journey and spread out the earth.
>
> *(7.87.1)*[6]

The double journey refers to the fact that in the Vedic cosmology the ecliptic was divided in half, into two *ayanas* or paths, the northern path, called *devayana*, and the southern path, called *pitryana*. The two paths were often contrasted: the *devayana* was the path of the *devas*, or gods; the *pitryana* the path of the *pitris*, or fathers (ancestors). These two paths divided the year between them, thirteen *naksatras* belonging to the *devayana*, thirteen to the *pitryana*, the remaining *naksatra* signaling the central day, *vishuvan*, which separated the two halves of the year.[7] The northern path, the *devayana*, was the path across the heavens taken by the sun when it moved to the north of our equinoctial marker, from the spring equinox to the autumn equinox. The southern path, the *pitryana,* comprised the other half of the year, from autumn to spring. [figure 2.3]

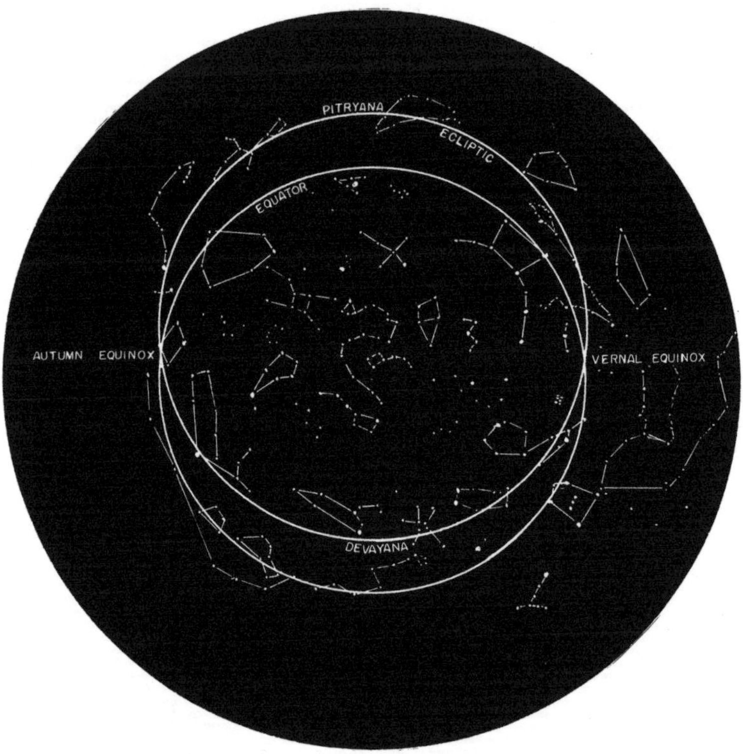

Figure 2.3
The Devayana and Pitryana

These two paths were imagined to be not only the paths of the sun and moon, but also paths for the souls of the dead. The *pitryana* was the path originally taken by Yama, the first man to reach the other world and thereby becoming the King of the Dead. The souls of the deceased follow this path, eventually reaching *yamaloka,* Yama's realm, the heaven of the ancestors. A funeral hymn in the Rig Veda(10.14.1,2,7,8,10,11)[8] describes the *pitryana:*

> The one who has passed beyond along the great, steep straits, spying out the path for many, the son of Vivasvan, the gatherer of men, King Yama—honor him with oblation.
> Yama was the first to find the way for us, this pasture that shall not be taken away. Where our ancient fathers passed beyond, there everyone who is born follows, each on his own path.
> *[To the dead man:]* Go forth, go forth on those ancient paths on which our ancient fathers passed beyond. There you shall see the two kings, Yama and Varuna, rejoicing in the sacrificial drink. Unite with the fathers, with Yama, with the rewards of your sacrifices and good deeds, in the highest heaven. Leaving behind all imperfections, go back home again; merge with a glorious body.
> Run on the right path, past the two brindled, four-eyed dogs, the sons of Sarama, and then approach the fathers, who are easy to reach and who rejoice at the same feast as Yama.
> Yama, give him over to your two guardian dogs, the four-eyed keepers of the path, who watch over men. O king, grant him happiness and health.

In India the *devayana* and *pitryana* eventually became the path of fire and the path of smoke, the path of fire leading to escape from the wheel of rebirth, the path of smoke leading to reincarnation. In this later development the original astronomical sense remains only as a half-buried vestige, as we can see in the Chandogya Upanishad (5.10.1-5):[9]

> Those who know this as well as those who worship in the forest knowing that self-mortification is the same as faith, merge into the flame (of the funeral pyre); from the flame (they pass on) into the day, from the day into the half-month of the full moon, from the half-month of the full moon into the six months during which the sun moves northwards. From [those] months into the year, from the year into the sun, from the sun into the moon, from the moon into lightning. There, there is a Person who is other than human. He leads them on the Brahman. This path is the "way of the gods."
> But those who in their villages lay great store by sacrifice, good works and the giving of alms, merge into smoke, from smoke [they pass on] into the night, from the night into the latter half of the month, from the latter half of the month into the six months in which the sun moves southwards. These do not reach the year. From [those] months they [merge] into the world of the ancestors, from the world of the ancestors into space, from space into the moon which is King Soma, the food of the gods. This the gods eat up. There they remain until the residue [of their good works] is exhausted, and then they once again return on the same path.

I have dwelt at some length on this ancient notion of the ecliptic divided into two paths at the equinoxes because these two paths underlie the very framework of the Odyssey. In the Vedas the *devayana,* also called the *uttarayana* (meaning the higher, or northern, or left path, since when facing east the north is on the left) was the white path; the pitryana, also called the *daksinayana* (meaning the southern, or right-hand path) was the black. By Homer's time the contrast had undergone a sea-change, becoming richer and more subtle: the domestic world of Ithaca, governed by Athene, is played against the fantastic world of Odysseus's wanderings, governed by Poseidon. As we shall see in the following chapters, the wheel of Odysseus's wanderings pictured in Chapter One, even though it resembles the wheel of twelve signs of the zodiac familiar to astrologers, covers only half of the ecliptic, the thirteen *naksatras* lying along the path of the sun between the spring equinox and the autumn equinox, each adventure corresponding to a naksatra in a congruent sequence. The voyage of Odysseus, in other words, is a passage along the northern path, the *devayana.* And the homecoming of Odysseus can be read in the stars that mark the other path of the sun and moon, the *pitryana,* from the autumn to the Spring equinox.

One curious thing about this ancient, half-faded pattern in the weave of the Odyssey is that it seems to be reversed from the Vedic conception of the two *ayanas*. We should expect Odysseus's voyage to follow the *pitryana*, since in the course of his journey he visits the realm of the departed ancestors, the pitris; instead we find this realm on the northern path of the sun, the *devayana*. The Vedas exhibit this same discrepancy. For example, even though the *pitryana* is described as the path to Yama's realm, the passage in the Rig Veda quoted earlier advises the departed soul to

> run on the right path, past the two brindled, four-eyed dogs, the sons of Sarama, and then approach the fathers, who are easy to reach and who rejoice at the same feast as Yama.
>
> *(10.14.10)*

The dogs that guard the path to the land of the dead reappear in the mythologies of Iran, Greece, and Scandinavia, and are still to be seen in the night sky. They are the two dogs that follow on the heels of Orion, and are located, not on the *pitryana* as one would expect, but on the *devayana*. As we have seen, these two paths are defined in the Vedas with reference to the sun. But the very fact that the 27 *naksatras* exist at all points to an earlier lunar reckoning of time, and in this earlier reckoning the two paths were the reverse of what they later became. The full moon is always opposite the sun,

180 degrees away. When the sun enters the constellations that mark the autumn and the season of death, the full moon can be seen in the vicinity of the two dogs, indicating that the *pitryana* was at one time defined as the path of the full moon north of the equatorial marker, not the path of the sun. When the Aryans shifted to a solar calendar it is likely that the two *ayanas* were reversed to keep the correlation with the seasons.

Some of the correspondences between the Vedic list of *naksatras* and the sequence of Odysseus's adventures are obvious, some not so much so, which is to be expected from a comparison of these two descendants of a very ancient tradition, especially when the descendants found their way into print 3000 miles apart and perhaps 2000 years distant from their obscure ancestor. But even though some portions of the two texts have become obscured by time, the pattern endures, surviving beneath the Mycenaean trappings of the Odyssey like a palimpsest, discernible to anyone who is willing to grant that those who lived before the dawn of history may not have lived in total darkness.

Before we chart the path of Odysseus's voyage through the stars between the spring and autumn equinoxes, one problem remains in our own path: where in the heavens can we find the equinoxes? One might expect that the precise point on the sun's path where it crosses from the *devayana* to the *pitriyana* (three o'clock on the celestial clock, as it were) would be a simple matter to locate. And so it would be, if it weren't for the annoying fact that the face on the clock doesn't stand still.

Suppose we could hold a piece of chalk against the bowl-shaped blackboard of the sky as it makes a 24-hour revolution, and suppose that chalk was held over our central post, the one pointing due east. We would be able to trace a great circle in the heavens, and this circle would be the celestial equator, dividing in half the imaginary sphere of stars that surrounds us. At the present time we can watch the sphere of stars rotate about the north star, Polaris. And we must emphasize "at the present time," because this center of rotation slowly changes its position over time, describing a circle of its own in the northern sky.[figure 2.4]

figure 2.4
The Migration of the Celestial Pole

As mentioned earlier, the pole was near the star Kochab 3000 years ago, and nearly 5000 years ago it was located at the star Thuban in the constellation Draco. Now if this center, marking the axis of rotation of the celestial sphere, moves to a different place among the stars, so must the great circle of the celestial equator which divides the sphere in half. At the present time the equator runs through the three stars that make up the belt of Orion. Another way to say this would be that these three stars can be seen to rise directly over our central post on the eastern horizon. Six thousand years ago an astronomer would have seen Orion's shoulders rise over the central post and the belt rise well to the south of it.

The moments when the sun crosses from the *devayana* to the *pitryana* or the reverse, the equinoctial moments, can be located in space, against the background of the stars. At the present time the spring equinox occurs in the constellation Pisces, the autumn equinox in Virgo. These constellations now mark the points at which the ecliptic crosses the celestial equator. If the path of the sun and moon, the ecliptic, remains constant, as it nearly does, but the equator changes its position relative to the fixed stars, the points at which the sun crosses the equator will move. The movement is extremely slow, so slow that historians of astronomy are reluctant to believe that it was perceived earlier than the second century B.C., when Hipparchus is said to have discovered the movement, called the "precession of the equinoxes." Hipparchus was the first to work out in mathematical terms the pace of the precession; however, it is unlikely it went unnoticed. It could have been observed by anyone blessed with curiosity, a long life, and an oak tree to sight across.

If, for example, we lived in that far-off time when the heliacal rising of the Pleiades announced the equinox, we might have noticed that the sun had gained on the Pleiades between our tenth and our eightieth year—the shift is one degree, or about twice the apparent diameter of the moon, in 72 years. If the Pleiades first faded with the equinoctial sunrise while still fairly high over the eastern horizon, the constellation would have served to usher in the equinox over the course of several hundred years. Eventually, however, with the sun at the equinox creeping closer and closer to the Pleiades, the constellation would not appear at all over the eastern horizon just before the sun rose over our central marker. It would have been seen very faintly in the dawn's early light the day before, and perhaps we might advance the day of the equinox (March 20 instead of March 21, to give a current example) to prolong the Pleiades' usefulness for another 70 years. But obviously we couldn't get away with such fudging indefinitely, especially if we happened to be Indo-Europeans living north of the forty-fifth parallel, where around the time of the equinox we would observe on successive days the rising sun shift its position on the horizon a distance equivalent to its own diameter. Eventually the alignments would be so far off that the Pleiades would have to be abandoned, and the next asterism in line, the *naksatra* called *Apabharanis* ("the bearers"), would be designated as the herald of the equinox. The same shift would be required for the other three corners of the year—the autumn equinox and the winter and summer solstices.

If the tradition of careful astronomical observation extends as far back in time as it must, given the correspondences between the Hindu Vedas and the Homeric Odyssey, that is, before the Indo-Europeans became Indians and Europeans, we can expect that some evidence of this shifting equinoctial point still lingers in the astronomical poem so religiously memorized and passed from generation to generation. In the next chapter I hope to demonstrate that such a residue survives in the Odyssey, testifying to a tradition of astronomical observation that began as early as 4000 B.C. and lasted for 3000 years.

CHAPTER THREE

✯ *The Gates of Heaven*

Whereas other animals hang their heads and
look at the ground, he made man stand erect,
bidding him look up to heaven,
and lift his head to the stars.
(Ovid, *Metamorphoses, Bk I)*

Any harbinger of the vernal equinox, that is, any star that would herald the sun's appearance over our central marker in Homer's time and for thousands of years before, will be found in what is called the Eastern Quarter of the heavens. This quarter contains five stars of the first magnitude or brighter—Sirius, the dog star; Rigel and Betelgeuze in the constellation Orion; Capella in Auriga; and Aldebaran in the Hyades. To the west of Aldebaran lies the distinctive cluster of stars known as the Pleiades, often appearing as a blur unless the air is especially clear, at which time six stars can be made out easily. Further to the west lie the four stars that make up the constellation we know as Aries, the Ram. Nowadays these are called winter constellations, since they are visible in the southern sky after sunset from November through February[figure 3.1].

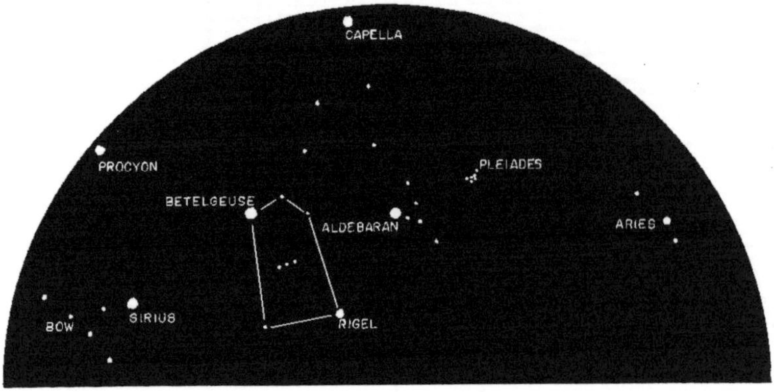

Figure 3.1
The Eastern Quarter

The Indo-Europeans, getting up before dawn to tend to their cattle, saw them rise heliacally in the spring.

Indo-European names still cling to these stars. The Greeks adopted many Babylonian constellations in the centuries after Homer. Scorpio, Capricorn, Sagittarius, Pisces, Leo, Taurus, and others are Babylonian in origin. But the words Sirius, Orion, Hyades, Pleiades, and Aries are neither Babylonian nor Egyptian. They have Indo-European roots, roots that extend so far back in time that the original forms of these words and their attendant meanings are extremely obscure. We can, however, be fairly certain about the reason for their continued importance to the Indo-Europeans: except for Sirius, these constellations took their turns signalling the spring equinox and the beginning of the year over a period of 3000 years, from around 4000 B.C. to the Greek Dark Age.

This quarter is dominated by Orion, the largest and certainly the most striking of all the constellations, and probably the easiest to make out in the night sky. His knees and shoulders are marked by a rectangle of bright stars, three stars in a line make up his belt, and another line of stars are usually imagined as a sword hanging from the belt. Orion is a giant, a mighty hunter, often pictured with a club or a hunting net in his left hand. His dogs, Sirius and Procyon, the two dog stars, follow at his heels. No adequate etymology of his name has ever been uncovered, and scholars can only conclude that the name is pre-Greek.[1]

Lying on the celestial equator, Orion can be seen from every part of the globe, and he figures in myths and legends all over the world. From his size and beauty, we might expect stories of heroic exploits, exploits that ought to rival those of Hercules. The myths speak otherwise. Homer gives us a typical one, told by Calypso when she was ordered to let Odysseus continue on his way:

> A cruel folk you are, unmatched for jealousy, you gods who cannot bear to let a goddess sleep with a man, even if it is done without concealment and she has chosen him as her lawful consort. You were the same when Rose-fingered Dawn fell in love with Orion. Easy livers yourselves, you were outraged at her conduct, and in the end chaste Artemis rose from her golden throne, attacked him in Ortygia with her gentle darts and left him dead. (5. 118-124)[2]

In another version of this tale, Apollo, fearful that Artemis might fall in love with Orion, tricked her into killing him. One day when Orion was swimming in the sea close to Ortygia Apollo challenged Artemis to a show of marksmanship, telling her to hit a distant object he showed her floating in the water. Artemis took the challenge, let fly an arrow, and Orion was pierced through the head.[3]

A third version of this same story tells how Artemis, outraged that Orion was pursuing her virgin companions, the Pleiades, sent a scorpion to sting him to death. A fourth claims that Artemis was defending her own virtue and not that of her companions when she sent the scorpion.[4] Other accounts say that the scorpion was sent by Hera, who devised this most humiliating of deaths to counter Orion's boast that he feared no animal on earth.[5]

The stories of death by a scorpion's sting are of a more recent origin than the others, and are often quoted to demonstrate that myths are often disguised accounts of natural phenomena, in this case the coincidence that the constellation Scorpio rises in the east just as Orion sets in the west. The myth probably dates from the Hellenistic age, after the Greeks had borrowed the zodiac, in which the scorpion figures prominently, from the Babylonians and Egyptians. In fact, the tale is closely related to the Egyptian myth of Horus, son of Osiris and Isis, who was killed by a scorpion sent by the god Set. Yet all the stories, whether early or late, have the same theme: Orion, because he was beautiful, or amorous, or boastful, had to die.

There is one ancient tale with a happier outcome. It seems that Orion fell in love with Merope (one of the Pleiades), the daughter of Oenopian. Oenopian promised that Orion could marry Merope once he rid their island of all the wild beasts that infested it. Orion did just that, but the father still refused to give up his daughter. In disgust, Orion got drunk, and then forced Merope to lie with him. Oenopian, in revenge, put out both Orion's eyes and turned him loose on the seashore. Having been told by an oracle that he could regain his sight if he traveled to the east and looked with his empty eye-sockets at the point where the sun first rises from Ocean, Orion set out to sea in a small boat. He followed the sound of a Cyclops' hammer, and reached the forge of Hephaestus. Putting one of Hephaestus's helpers on his shoulders to act as guide, Orion set out again to find the rising sun. They succeeded, and Orion's sight was restored.[6]

Between and slightly above Betelgeuse and Bellatrix, the two stars that mark the shoulders of Orion, lies a cluster of three faint stars. This cluster forms the third *naksatra* in the traditional list. It goes by several different names in the Vedic scriptures, among them *andhaka*, meaning "blind," and *mrgasiras,* meaning "the head of a stag." Scholars usually attribute the first of these names to the fact that this cluster of stars is quite dim. Whether the name gave rise to the myth of Orion's blinding or the myth gave rise to the name is impossible to say. It does seem, however, that the Greek story of the blinding of Orion goes back a very long way.

The other name of the *naksatra, mrgasiras*, "the head of a stag," is associated with a story of lust and vengeance similar to the affair between Orion and the Dawn told by Calypso. It is a well-known myth, and its connection with the constellation Orion is undisputed. The Aitareya Brahmana tells how Prajapati, Lord of Created Beings, committed incest with his daughter, Usas, the Dawn. Rudra, the Prince of Demons (later to become the god Shiva) became outraged at the crime. Now Rudra was a formidable archer, one whose arrows never missed. In order to escape him, Prajapati turned himself into a stag. The ruse didn't work, and he was brought down by Rudra's "three-jointed arrow." In other versions Usas turned herself into a red doe to escape Prajapati, and he in pursuit turned himself into a stag, but failed to out-run Rudra's arrow. These figures can be seen moving across the night sky as Orion, the stag, pursuing Usas, the red star Aldebaran, and in turn pursued by Rudra, the bright star Sirius,[7] whose three-jointed arrow is still lodged in the great stag's body.

The two stories, Rudra killing Prajapati and Artemis killing Orion, come together in the story of another famous hunter, Actaeon, whose name, meaning "shore dweller," calls to mind the blind Orion wandering along the shore (the shore being that of the great celestial river, the Milky Way). Out hunting with his dogs one day, Actaeon chanced upon Artemis bathing with her nymphs in a pool. Artemis, the virgin goddess, became furious that a mortal should see her naked. She splashed water in Actaeon's face, and he was transformed into a stag. Pursued by his own dogs, he was torn to shreds.

Every story about Orion that survives turns out to be a story about his destruction. Oenopian blinds him, Artemis kills him, Hera kills him, Rudra kills him, his own dogs devour him. Why should this be? Why this persistent story about the largest and most beautiful constellation in the heavens? What lies behind these myths, all of which say, in variation after variation, that Orion died?

The most likely answer is that Orion once marked the vernal equinox, and probably marked it when the Indo-Europeans first sorted out the heavens and the movements of the sun and the moon. If this is so, the time would have to have been somewhere around 4000 B.C. At that time Orion stood at the junction where three paths met—the *devayana*, the *pitryana,* and the Milky Way, and the three paths met again at the autumn equinox in the tail of the constellation we know as Scorpio. This coincidence shaped the spectacle of the heavens into a meaningful symmetry, one that has not been seen since, and one that will not be seen again for another 7000 years. Part of that meaning had to do with the fact that the Milky Way joined the equinoxes together. The Milky Way arched across the heavens and disappeared below the horizon, the arch revolving slowly throughout the night, keeping the two poles of the year together in the mind, if not in view. spring and fall, birth and death, heaven and earth would have all been bound together. Orion stood at the crossroads, guarding the threshold between winter and summer, the coming of life and its passing. The sun standing in Orion announced the beginning of the year and the season of rebirth, the full moon in Orion brought in autumn and the season of death. Thus the Milky Way became associated with that Other World, the Land of the Dead, sometimes as a river one must cross over, sometimes as a celestial path one must follow to reach that Land.

Unknown to the ancient astronomers, the precession of the equinoxes was slowly twisting their cosmos askew. As the centuries came and went Orion stood lower and lower on the eastern horizon at the moment of equinox. First his legs no longer showed, then his belt sank beneath the horizon, and finally there came a year when he did not appear at all on the appointed day. The keystone of a whole cosmology, the figure that held it together, was gone. What survived were tales upon tales about his death.

Three thousand years is a very long time for any tale to survive in the memories of a people, and yet it shows up again, thinly disguised, in the Odyssey of Homer. Let us join Odysseus as he explores the island of Circe:

> When the sun was up I took my spear and sword, slipped away from the ship, and struck inland, making for a coign of vantage from which I might look out for signs of human industry or hear men's voices. I climbed a rocky height which promised a wide view, and on reaching the top I was able to see the smoke rising from the distant spot where Circe's house lay screened by the dense oak-scrub and forest trees. That glimpse I had of ruddy smoke left me in two minds whether or not to press forward and reconnoitre. After some hesitation I thought the better course would be to return first to my ship on the beach, give my men a meal, and send out an exploring party. And here some god must have been moved to pity by my forlorn condition. For when I

> had almost got back to the ship, I fell in with a great antlered stag, right across my path. The fierce heat of the sun had brought him down from the forest pastures to drink at a stream, and as he came up from the water I caught him on the spine half-way down his back. The bronze point of my spear went right through him, and with a groan he fell in the dust and was dead. With one foot on his carcass I dragged the spear out of the wound, laid it on the ground, and left it there while I plucked myself some withes and willow-twigs, which I twisted into a fathom's length of rope, carefully plaited from end to end. With this I tied his feet together, and since he was far too big for me to carry on one shoulder and steady with a single hand, I slung the great beast around my neck, and using my spear as a staff I set off for my ship. (10.144-171)

Every Homeric scholar writing in the past hundred years hurries by the slain body of the magnificent stag without a word, anxious to arrive at Circe's palace in order to comment on the penned-up pigs. But when we come across what seems to be a complex of motifs that go back to a time before the Indo-Europeans began their migrations it might be wise to pause at the slain body of the stag for a moment. The first thing we notice is that the incident seems like a gratuitous intrusion into the narrative flow, an ancient, moss-covered boulder sticking up out of the stream. There seems to be no need for it, and yet the oral tradition has preserved it. There are many such antique stones lying in the path of Odysseus, and, as we shall see, many of them stand out from the narrative by their simple incongruity, their imperviousness to most "earthly" explanations, or their seemingly gratuitous presence. And in these we see one of the characteristics of oral tradition: it adds to itself, accumulating details as the culture changes, but it seldom forgets. This is borne out by the account of a curious encounter Odysseus has the next morning.

After an evening feasting on venison, Odysseus heads for Circe's palace again to look for his crewmen who have failed to return. On the path he meets the god Hermes, who gives Odysseus a strange plant as protection against Circe's powers of enchantment. The gods, we are told, call the plant *mōly*. We are never told how it is to be used (eaten? worn around the neck like garlic?), only that its root is black, its flower is milk-white, and only gods can dig it up. Botanical lore has been ransacked to identify this antidote to Circe's witchcraft, and wild rue, wild cyclamen, and, of course, garlic have all been suggested. But the real reason mortals can't dig it up is that it exists in the sky. *Mōly* is a derivative of *mula*, a Sanskrit word meaning "root." *Mula* is also the name of one of the Vedic *naksatras*, the group of stars in the tail of the constellation we know as Scorpio. As we have already seen, Scorpio is often associated with Orion. The black root and white flower of the plant call to mind the white and black paths of Vedic astronomy, the *devayana* and the *pitryana*, and when the *naksatra mrgasiras*, the head of the stag, signaled the vernal equinox, it was the *naksatra mula* that signaled the autumn equinox, the other place in the heavens where the two paths came together.

Now that we have located the unfortunate stag and the plant called *mōly*, let us follow Odysseus to Circe's palace. Perhaps she can lead us to a living equinox. She is not far away, but to find her we will have to dig her palace out from under several layers of cultural change laid down by more recent ages.

Above and to the west of Orion lies the constellation we know as Taurus, the Bull. Taurus is a Babylonian constellation, and has been superimposed on an earlier Greek constellation known to Homer and Hesiod as the Hyades, a group of six stars that shine like misplaced jewels on the muzzle of the Bull. The word "Hyades" derives from *us*, "pig," and *ades*, "many."[8] It is an ancient Indo-European designation, turning up among the Anglo-Saxons as well. They called the constellation "the Boar-throng."[9] The brightest star in the constellation of the Hyades is a red star called Aldebaran (from Arabic *Al Dabaran*, "the follower," i.e., following the Pleiades in its rising), popularly known as "the Bull's Eye." In the Vedas the star is known as *rohini,* meaning "red," and, as previously mentioned, is identified with the Dawn, Usas, who transformed herself into a red doe while fleeing from her father Prajapati. "Usas," incidentally, derives from the proto-Indo-European root **aues,* from which also derives the Greek word for dawn, *eos.* The Anglo-Saxon goddess of the Dawn, whose rites were celebrated at the vernal equinox, was called *Eostre*, the source of our "Easter."

So the original meaning of Hyades was "a group of pigs," and the brightest star in the constellation was known as the Dawn. Homer shows us Circe, daughter of the Sun, surrounded by pigs, those unhappy men in Odysseus's command whom Circe transformed, and tells us that Aeaea, Circe's island, is the home of the Dawn (12.3). But how can any place on earth or in heaven be called the Island of the Dawn? This question has perplexed those scholars who have looked for Odysseus's ports of call in the Mediterranean. Obviously, the further east you go, the more east there is, and no Island of the Dawn seems forthcoming, even though many scholars have settled on Circeo on the west coast of Italy, and that not even an island. The question might perplex us as well, since all the stars on the great circle of the ecliptic take their turn announcing the dawn in the course of a year, and any of them could deserve the name "Island of the Dawn." But there is one significant dawn, the dawn of the year, when the sun moves from the *ayana* of night to the *ayana* of day—the vernal equinox. After the death of

Orion, Aldebaran, or *rohini*, was the star that announced the equinox.[10] This is why Usas, the Vedic dawn, is identified with that star, and why Circe is said to live where Dawn has her dancing-floor.

This is simple and straightforward enough to look like a solution, and it would be if it weren't for the Laestrygonians. But they, too, lay claim to the equinox and, apparently, to the same constellation. Some time before Odysseus and his men beached their ship on Circe's island they had watched from outside the harbor of the Laestrygonians as the other eleven ships in Odysseus's squadron were pelted by boulders and the men in them harpooned and devoured by the local cannibals. The place is called "Telepylus," an unusual name, yet fitting for such a remarkable site:

> We came to Telepylus, the citadel of Lamos in Laestrygonia, where herdsman calls to herdsman as one brings in his droves and the other answers as he drives out his. There an unsleeping man could have got double wages, one wage for tending cattle and another for pasturing white sheep; for the paths of day and night are close together. (10.81-86)[11]

The term "Telepylus" is a compound: *tele* + *pylus* = far + gate. It is an odd word, occuring only this once in Homer, and a puzzle to translators.[12] The phrase "the citadel of Lamos in Laestrygonia whose gates are far away" makes little sense. But "far-gated" is an appropriate term for the Gate of Heaven, the point in the heavens where "the paths of day and night are close together," in fact where the *devayana* and *pitryana* come together at the equinox. The Shatapatha Brahmana states, "the day is of the Devas; the night of the Pitris" (ii. 1.3, 1-3).[13]

The Gate of Heaven is the entrance to the Land of the Dead and the Halls of Hades and Persephone. Hesiod tells us that Atlas once stood at this threshold, and his account, much expanded, is remarkably similar to Homer's:

> And here stand the terrible houses of dark Night, and the buildings are sheathed in the dark of the clouds. Before them Atlas, son of Iapetos, stands staunchly upholding the wide heaven upon his head and with arms unwearying sustains it, there where Night and Day come close to each other and speak a word of greeting and cross on the great threshold of bronze...
>
> (Theogony, 744ff.)[14]

When we think of Atlas we picture him as holding the globe of the earth on his shoulders. But this image is the last in a long series of misunderstandings. There is no indication in this passage that Hesiod thought of the earth as spherical. However, the fact that he has Atlas holding up the sky gave rise in the late first century B.C. to the image of Atlas holding on his shoulders a celestial globe, a spherical map of the stars.[15] Eventually this sphere was transformed into the globe of the earth. But originally Atlas supported, not our earth, nor an earlier celestial globe, nor even the sky, as Hesiod has it. Homer preserves the original image:

> Atlas . . . who holds up by himself
> The enormous pillars that hold earth and heaven apart
> (1.52-54)

These pillars stand on either side of Hesiod's bronze threshold, and form the gate whose name, according to Homer, is Telepylus.

Before we move on to other constellations lying beneath the surface of the Odyssey, one last tale about Orion is worth re-telling. The tale can be found in the Bible among the accounts of the protracted war between Israel and the Philistines, those Indo-

Europeans who invaded Palestine sometime after 1200 B.C. This time Orion is not called Prajapati, or Actaeon, but Samson. In the Bible he is known as one of the Judges of Israel; however, his identification with the constellation Orion is clearly established by the strange weapon he uses to slaughter the Philistines:

> When he came to Lehi, the Philistines came shouting to meet him; and the Spirit of the Lord came mightily upon him, and the ropes which were on his arms became as flax that has caught fire, and his bonds melted off his hands. And he found a fresh jawbone of an ass, and put out his hand and seized it, and with it he slew a thousand men. And Samson said, "With the jawbone of an ass, heaps upon heaps, with the jawbone of an ass have I slain a thousand men." When he had finished speaking, he threw away the jawbone out of his hand, and that place was called Ramathlehi [i.e., the hill of the jawbone]. And he was very thirsty, and he called on the Lord and said, "Thou hast granted this great deliverance by the hand of thy servant; and shall I now die of thirst, and fall into the hands of the uncircumcised?" And God split open the hollow place that is at Lehi, and there came water from it; and when he drank, his spirit returned, and he revived. Therefore the name of it was called En-hakkore [i.e., the spring of him who called]; it is at Lehi to this day. And he judged Israel in the days of the Philistines twenty years.
>
> (Judges 15:14-20 RSV)

One would have to agree that this Judge was no ordinary mortal. Neither was the jawbone any ordinary jawbone. It is visible today in the night sky above the shoulder of Orion, still lying where Samson threw it. It goes by the name of the Hyades. Three thousand years ago it bore the name "Jaw of the Bull" among the Babylonians.[16] As we mentioned before, the Babylonian constellation Taurus, the Bull, coincides with the Indo-European Hyades. And the Hyades do form the V-shape of a jawbone.

Homer tells us that outside the village of the Laestrygonians is a bubbling spring called *Artacie* (10.108) This spring seems to have been a fixture in ancient lore, for Jason and the Argonauts also came across it (*Argonautica* 1.957). It has long been suspected that the voyage of the Argonauts took place among the stars (Isaac Newton may have been the first to bring the matter up), although as yet no detailed and coherent analysis of their celestial adventures has been published. In the *Argonautica* of Apollonius the spring called *Artacie* was to be found near a fine harbor, the harbor of Chytus called "Fairhaven," where savages hurled rocks down on Jason's ship. The harbor of the Laestrygonians is described by Homer as a narrow entrance flanked by towering rocks (10.87-90), and these Laestrygonians also hurled boulders down on the fleet of Odysseus. Odysseus's fleet was lost; Jason's men succeeded in overcoming the savages.

It takes no great stretch of the imagination to see that Homer and Apollonius are probably talking about the same place. It may not be so easy to accept that the spring in the Odyssey called Artacie and the spring in the Old Testament called *En-hakkore* are the same spring. Perhaps the difficulty can be alleviated by recognizing that Samson and Atlas are cousins, and not such distant cousins either, both embracing the pillars that hold the sky in place:

> When Delilah saw that he had told her all his mind, she sent and called the lords of the Philistines, saying, "Come up this once, for he has told me all his mind." Then the lords of the Philistines came up to her, and brought the money in their hands. She made him sleep upon her knees; and she called a man, and had him shave off the seven locks of his head. Then she began to torment him, and his strength left him. And she said, "The Philistines are upon you, Samson!" And he awoke from his sleep, and said, "I will go out as at other times, and shake myself free." And he did not know

> that the Lord had left him. And the Philistines seized him and gouged out his eyes, and brought him down to Gaza, and bound him with bronze fetters; and he ground at the mill in the prison. But the hair of his head began to grow again after it had been shaved.
>
> Now the lords of the Philistines gathered to offer a great sacrifice to Dagon their god, and to rejoice; for they said, "Our god has given our enemy into our hand, the ravager of our country, who has slain many of us." And when their hearts were merry, they said, "Call Samson, that he may make sport for us." So they called Samson out of the prison, and he made sport before them. They made him stand between the pillars; and Samson said to the lad who held him by the hand, "Let me feel the pillars on which the house rests, that I may lean against them." Now the house was full of men and women; all the lords of the Philistines were there, and on the roof there were about three thousand men and women, who looked on while Samson made sport.
>
> Then Samson called to the Lord and said, "O Lord God, remember me, I pray thee, and strengthen me, I pray thee, only this once, O God, that I may be avenged upon the Philistines for one of my two eyes." And Samson grasped the two middle pillars upon which the house rested, and he leaned his weight upon them, his right hand on the one and his left hand on the other. And Samson said, "Let me die with the Philistines." Then he bowed with all his might, and the house fell upon the lords and upon all the people that were in it.
>
> (Judges 16:18-30 RSV)

The mill that Samson is compelled to turn is an old one, left over from an ancient cosmology that predates the conflict with the Philistines by several thousand years. The millstone is, in fact, the heaven itself, forever turning above our heads. It is found in the ancient lore of peoples from Persia to Iceland. In Norse mythology the mill, named Grotte, the crusher, was owned by Frodhi, who recruited two giant maidens, Fenja and Menja, to work it night and day. The very same mill appears in Apollonius's account of the Voyage of the Argonauts (iv. 1090-95) where it is turned by a daughter of the ogre-king Echetus, who blinded her with brazen spikes. The mill emerges again in the Odyssey (20.105-120)[17] where it is turned by one of the women servants in the household of Penelope. In the Kalevala, the great Finnish epic, the mill is called the Sampo, a word derived from the Sanskrit *skambha* which, as we saw in the previous chapter, denotes a pillar or pole, referring to the World Axis.[18] The World Axis culminates in the Pole star, around which the heavens revolve. In about 1000 B.C. when the star beta Ursa Minoris was the Pole Star, this star was named Kochab in Persia, "the mill peg," around which the mill of heaven rotated. The Arabs called the pole star Kotb, the axle. They, too, envisioned the bowl of heaven as a turning millstone, and the celestial pole as the axle bearing in which the mill iron turns.[19]

So we have the blind Samson, alias the blind Orion, turning the Mill of Heaven. Samson at the pillars that hold up the roof is identical to Atlas holding the pillars that hold earth and heaven apart in the passage from the Odyssey quoted above, pillars which form the Gate of Heaven. This Gate is the entrance to the Land of the Dead, also known as Telepylus, the place where the paths of day and night, the *devayana* and *pitryana*, meet, the vernal equinox, the gate of the year. As in the other tales of Orion, Samson must die, and at his death the Gate of Heaven is destroyed. Should one object that the pillars holding up the roof of the house of the Philistines are nothing like a gate, we can point to another version of this strange tale in the Book of Judges (16.3):

> "Samson lay til midnight, and at midnight he arose and took hold of the doors of the gate of the city and the two posts, and pulled them up, bar and all . . ."

After the death of Orion, the next star to serve as the equinoctial herald was Aldebaran, followed 800 years later by the Pleiades, which were in turn succeeded by a star in the constellation Aries. With each succeeding age the intricate tale used to keep track of the celestial phenomena would have had to be recast. This was not as simple as burning the old book and writing a new one. As these tales about Orion demonstrate, the oral tradition does not discard nor intentionally forget, though it may accumulate details with each succeeding generation. Given the conservative nature of the oral tradition we can expect a residue of the former cosmological structure to carry over into each new age. We have already seen this with regard to the description of Circe's Island of the Dawn, signifying the equinoctial point, and with Telepylus, also a name for the equinox, used in connection with the Laestrygonians. In the next chapter we shall follow the progress of the precession as it moves across the next two *naksatras*, leaving a residue of its passage in the descriptions of Aeolia and the cave of the Cyclops.

CHAPTER FOUR

✲ *Further Remnants of the Procession*

Many a night I saw the Pleiads, rising thro' the mellow shade,
Glitter like a swarm of fireflies tangled in a silver braid.

(Tennyson)

Before encountering the Laestrygonians Odysseus spends a month on the floating island of Aeolia being entertained by Aeolus and his family. And it is a bizarre family, to say the least. Aeolus has given his six sons to his six daughters in marriage, and at night the sons sleep by their loving wives in well-made wooden beds. I, and I think most of us, find the scene both charming and appalling. Odysseus is simply charmed, and has only words of praise for the luxury of the palace and the hospitality of his host. Not a word of dismay or even surprise at this thicket of incest. Perhaps, somewhere back in the early stages of the poem, long before it made its way to Greece and the ears of Homer, there were once reasons for imagining this disgusting family.

The floating island of Aeolia has been a great nuisance to those who have tried to reconstruct a geography of the Odyssey. Homer tells us it has cliffs that rise sheer from the sea, and lies a ten-days' sail due east of Ithaca. No island of that description has been found, and no island in the Mediterranean in legend or tradition has ever been known as the home of the winds. "You will be able to locate the wanderings of Odysseus when you find the cobbler who sewed up the bag of winds," said Eratosthenes.[1] The cobbler has disappeared into the darkness of pre-history, but traces of the bag remain in the night sky, and in the same place as the winds and the children of Aeolus. The place is the Pleiades.

The Pleiades are a cluster of six stars lying nearly on the ecliptic, to the west of Aldebaran, the star of the Dawn. They, like the Hyades, have been forced to share their heavenly home with Taurus, and are variously located on his shoulder, back, hind quarters, or tail, depending on how the figure of the Bull has been imagined. Six stars can be seen in the cluster, yet the memory of a seventh has lingered in traditional stories of the Pleiades. One explanation holds that one of the stars lost some of its brilliance within the memory of mankind and can no longer be seen, at least with the naked eye; another holds that there never were seven, this number having been attributed to the Pleiades by the Babylonians simply because the number seven played a significant role in their cosmology. The Vedic priests counted only six, leaving us at sixes and sevens over this little mystery. Whatever the reason for the lost Pleiad, this cluster of stars has been imagined the world over as a group of maidens.[2] Hesiod calls them "the virgin stars," the daughters of Atlas.[3] By the time Ovid wrote about them they had been deflowered by Zeus, Poseidon, Ares, and Sisyphus.[4] So it goes.

As mentioned earlier, the Pleiades are the first in the list of Vedic *naksatras*, indicating that they signaled the vernal equinox and the beginning of the year when this version of the list was formed. As such, the six stars could well have been imagined as the year itself, the twelve months paired off two by two. Evidence that the twelve months were indeed regarded as paired can be found in a complex hymn of the Rig Veda, sung in the form of a riddle:

> They say that besides those born in pairs there is a seventh born alone, while the six sets of twins are the sages born from the gods. The sacrifices for them are firmly set, but they change their forms and waver as he stands firm.
> They are female, but people tell me they are male. He who has eyes sees this, but the blind one does not understand. (1.164.15,16)

The translator of this passage, Wendy Doniger O'Flaherty, supplies this explanation: "In the ritual, there are twelve paired months and one odd one, the intercalary month which interrupts the sequence and causes the others to 'change and waver.' The androgyny of the creators appears, on the ritual level, in the fact that the months (a masculine term in Sanskrit) are procreative."[5]

Heraclitus believed that the twelve children of Aeolus were the twelve months of the year, reasoning that the months "marry," the six females representing the six months when the earth brings forth its fruits, the six males the months when the seeds are sown.[6] However, if the children of Aeolus do represent the twelve months, a less fanciful explanation of their unconventional behavior is that in Vedic times the year was divided into six seasons, each comprising two months.

The name "Pleiades" derives from the Indo-European root *pelu*, "many," the same root that gave birth to "polis" and eventually to our own "plenitude," "plethora," "plus," etc.[7] Scholars in antiquity assumed the word was an offspring of *plein*, "to sail," because the rising and setting of the Pleiades signaled the beginning and end of the season safe for sailing.[8] Hesiod gives us a vivid picture of the winds associated with the Pleiades:

> When the Pleiades, running to escape from Orion's grim bulk, duck themselves under the misty face of the water, at that time blasts of the winds are blowing from every direction, then is no time to keep your ships on the wine-blue water.
>
> Think of working your land instead, as I keep telling you.
>
> Haul your ship up on the dry land . . .
>
> (Works and Days, 619-624)[9]

The association of the Pleiades with the winds, though apparently a commonplace by the time of Homer, was probably not an ancient notion, as ancient notions go. At least as early as 2200 B.C., when the Pleiades last marked the vernal equinox and were the "mouth of the year," the Indo-Europeans apparently associated the Pleiades either with an animal skin or hide, or with a net to capture something with, or possibly both. The Vedic name for the Pleiades was *Krttikas*, a word whose roots are only partially clear. It may have derived from **krtti*, "to spin," or from **krta,* "net" or "web," or from *krtti,* "hide" or "skin."[10] Choosing the last of these etymologies would, of course, enhance my argument, especially since it is difficult to imagine the winds captured in a net, no matter how finely woven. Whichever one we choose, it does seem reasonable to connect the twelve incestuous children and the home of the winds with the Pleiades rather than with one of the Lipari Islands near Sicily, as so many geographers of the Odyssey have done.

The slaying of the stag and the plant called "*mōly*" on Circe's island, the paths of day and night coming together in the land of the Laestrygonians, the strange family of Aeolus—all these odd details point to particular stars in the eastern quarter of the heavens, stars which turn up in the Vedic list of houses of the moon. Moreover, each of these details carries with it evidence that the stars in question once served to announce the equinox and the new year.

Following the precession of the equinoxes still further across the dim centuries of pre-history we come to the *naksatra bharanis*, the equinoctial marker visible from 2200 B.C. until around 1200 B.C., the time of the Trojan War. *Bharanis* is a faint group of three stars (35, 39, 41 Arietis) in the constellation of Aries, the Ram. If the sequence of Odysseus's adventures mirrors the sequence of naksatras listed in the Vedas, it is here that we should expect to find the cave of the cyclops Polyphemus. When we discover that the Sanskrit *bharanis* means "the bearer" or "that which carries away"[11] what immediately comes to mind, of course, is the huge ram under which Odysseus makes his famous escape. The connection seems almost too obvious to mention; yet, unlike the other constellations mentioned so far—Orion, the Hyades, the Pleiades—the celestial figure of the Ram cannot be traced very far back in time with any degree of certainty. Of the twelve constellations that make up the zodiac it is the only one that is not Babylonian in origin,[12] and it is highly unlikely that it came from the Egyptians.[13] That leaves the Greeks and their ancestors. Neither Homer nor Hesiod explicitly mentions a constellation called the Ram, but this is no proof of its age one way or another. There is, however, one curious detail in the lore of the heavens that suggests the Greeks had known of a celestial ram for a very long time. Along with the Hyades and Pleiades, the Ram competed for space in the sky with Taurus, the

Bull. The Bull, already burdened with the Hyades on his nose and the Pleiades on his shoulders, apparently had to sacrifice his hindquarters to make room for this interloping Greek constellation. For this to happen the Ram must have been just as firmly established in Greek tradition as were the Hyades and the Pleiades, implying a much longer life than the written records reveal.[14]

Even though we might agree that the ram that bore Odysseus out of the cave can be identified with *bharanis*, we still have to ask, Where is the Cyclops? There is no configuration of stars in the immediate neighborhood of Aries that might resemble this giant. If we are to locate him in the sky at all, his gruesome blinding would seem to identify him, too, with the blind Orion.

Such is the case. Watching for the equinox, the astronomers waited for the heliacal setting of Orion. At the appointed time, the giant Orion could be seen sinking below the western horizon a few hours after sunset on the eve of the vernal equinox. His heliacal setting coincided with the heliacal rising of *bharanis*. Observing this coincidence was necessary, because the group of three stars that make up the *naksatra bharanis* are rather faint, and not easily seen close to the horizon. At dawn, the astronomers must have also used the first-magnitude star Capella, which, during the second millennium B.C. in the Greek latitudes, rose ahead of the sun at the same time as *bharanis*. Capella served to locate *bharanis* when the viewing conditions were poor.

Capella is the major star in the constellation Auriga, the charioteer, located above the constellation Orion. Auriga is as large as Orion, though less distinct, shaped like a crooked kite. Since the sixth century B.C. this constellation has been pictured as a young man with a whip, but no chariot, carrying a she-goat on his shoulder and two kids on his forearm. This strange figure is apparently the bastard child of two traditions. Like Taurus superimposed on the Hyades and Pleiades, this chariot driver, a Babylonian constellation, has been superimposed on the Indo-European constellation of a she-goat (the star Capella) and her kids.[15]

Before encountering the Cyclops, Odysseus and his fleet of twelve ships all run aground on an island, "not near yet not far" from the cave of the Cyclops. The island is over-run with wild goats. All the men in the squadron engage in a great hunt, and they slaughter 108 goats all told. Without Capella and her kids there would seem to be no reason for that strange and seemingly unnecessary interlude. Their presence in the night sky in the neighborhood of Orion explains why Odysseus's men fed on goat rather than venison or lamb.

As mentioned before, the astronomical information carried by the oral tradition is most often found in those gratuitous or incongruous details of Homer's narrative. Another such incongruity is in the description of the weapon used to put out the eye of Polyphemus:

> The Cyclops has a huge staff of green olive-wood, which he had cut to carry in his hand when it was seasoned. To us it looked like the mast of some black ship of twenty oars . . . I set to work and cut off a fathom's length. (9.319-325)

Anyone who has seen the olive trees of Greece might wonder at this description. Olive trees are renowned for their twisted shapes. In Greece it is said that an olive tree will twist completely around every hundred years. The last tree one would use to make a mast would be an olive, unless he wanted the mast to look like an enormous corkscrew. Homer's analogy does convey the enormous size of the giant's staff, and one only wishes he had chosen another tree rather than the olive. Such quibbles aside, the use of this weapon is a deviation from the usual story. In all other versions of the tale the giant roasts his victims over the fire, and the hero blinds the giant with the iron spit the giant had been using. Denys Page surmises that the olive stake was substituted for the red-hot iron spit because the spectacle of the giant roasting his victims before eating them was too barbaric for Homer's audience.[16] This explanation is hardly adequate, since eating the victims raw might be considered even more barbaric, and the delicate sensibilities of Homer's imagined audience certainly aren't spared in the gruesome description of the blinding, in which the eyeball hisses and crackles from the heat of the red-hot point of the olive stake. A more sensible explanation, though not a simpler one, is to connect this weapon with the other two olive trees in the Odyssey, both as peculiar as this one: the olive tree around which Odysseus had built his bed, and the curious double-olive that he hides under after washing ashore at Scherie. Homer tells us

> he crept under a pair of bushes, one an olive, the other
> a wild olive, which grew from the same stem
> (5.475-476)

This is a peculiar detail for the oral poets to have memorized and faithfully transmitted, and it has given rise to various scholarly disputes about the botany of olive trees. Perhaps the dispute will be settled by the news that the Vedic *naksatra* corresponding to our constellation Libra was called *visakha*, "possessed of two branches."[17] When *bharanis*, "that which carries away," signaled the vernal equinox, *visakha* signaled the autumn equinox. Our third olive tree, the one attached to Odysseus's bed, in turn marked the winter solstice. A more complete discussion of these celestial trees will be presented as we follow Odysseus on his journey through the heavens.

Bharanis marked the equinox for a thousand years, from about 2200 B.C. down to the Trojan War and the subsequent collapse of the Mycenaean world. Homer's Odyssey, even though written down some five hundred years later, does not record the precession of the equinox beyond the last possible date that *bharanis* served as the marker, around 1200 B.C., and we must assume that the astronomical tradition that had held together for millennia came to an end with the Greek Dark Age.[18] If the long-standing practice of keeping watch on the heavens had continued down to the time of Homer, we should expect the equinox to show up in the adventure prior to the encounter with the Cyclops, that of the Lotus-eaters. But Homer gives the Lotus-eaters the briefest of treatments, only twenty lines, and I can find nothing in these lines that suggests a stellar connection, let alone an equinoctial point.

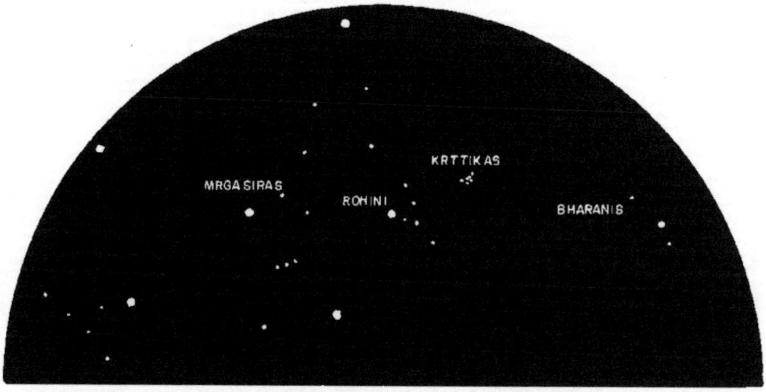

Figure 4.1

The Eastern Quarter Showing The Naksatras

The precession of the equinoxes moves backwards along the ecliptic, in the direction opposite to the movements of the sun and moon and planets. So far we have been following the trail left by the precession as it has moved across four *naksatras: mrgasiras* in Orion, *rohini* in the Hyades, *krttikas* in the Pleiades, and *bharanis* in Aries. [Figure 4.1] In its 3000-year history the poetic tradition carried on by astronomers conserved certain terms and images while the precession moved the equinox from one asterism to the next. As we have seen, this conserving led to some overlapping. The only way to sort out this overlapping of celestial elements in the Odyssey, it seems to me, is to trust the sequence that Homer and his predecessors took such pains to memorize. To put things the right way round and put Odysseus on his celestial path, we need to reverse the order of adventures discussed so far so that it is in keeping with the sequence given us by Homer. As table 4.1 shows, this sequence is in perfect accord with the sequence of *naksatras* given in the Vedas.

TABLE 4.1

Homer's term	*naksatra*	*meaning*	*asterism*
Cyclops	bharanis	the bearer away	35,39,41 Arietis
Aeolia	krttikas	skin or hide	Pleiades
Laestrygonia	rohini	ruddy	Aldebaran
Circe	mrgasiras	deer's head	$\lambda\ \phi^1\ \phi^2$ Orionis

CHAPTER FIVE

✧ *Okeanos and Other Celestial Rivers*

Then looking onward I made out a throng
assembled on the beach of a wide river,
whereupon I turned to him: "Master, I long

to know what souls these are, and what strange usage
makes them as eager to cross as they seem to be
in this infected light." At which the Sage:

"All this shall be made known to you when we stand
on the joyless beach of Acheron."

(Dante, *Inferno, Canto 3, Ciardi translation)*

Orion stands on the shore of the celestial river known as the Galaxy (from the Greek *gala* = milk) or Milky Way. Few of us in this latter half of the twentieth century have been privileged to see the Milky Way in its full glory; but to anyone fortunate enough to have lived before the inventive mind of Thomas Edison came along, the Milky Way was a common sight, and any mind, ancient or modern, could not fail to be awed by that river of a million stars. Homer mentions a few constellations—Orion, the Great Bear, Bootes, the Pleiades—leaving later generations to wonder why the most spectacular configuration in the night sky, covering one-tenth of the visible heavens and containing nine-tenths of the visible stars, should have escaped his notice. The truth is, Homer refers to the Milky Way frequently, but his meaning has gone unrecognized for centuries by men who could see in Homer only a primitive or fanciful geography, when in actuality Homer was describing a universe. Homer called the Milky Way *Okeanos*, usually translated as "River of Ocean" or "Ocean-stream."

It is commonly believed that the Greeks of the Homeric age imagined Okeanos to be a circular river surrounding a flat, disk-shaped earth. This mistaken belief seems to have taken root as early as the fifth century B.C., and was as puzzling to the thoughtful then as it is today. Herodotus was doubtful: "They [the Greeks] affirm in words that Okeanos, beginning from the sun's rising, flows round the whole earth, but they give no effective demonstration of this."[1] Aristotle, unconvinced, suggested that what the ancients meant by "Okeanos" was the moist vapor, in which clouds were formed, that encircled the earth.[2]

Homer never stated explicitly that Okeanos was a circular river surrounding the flat earth. Commentators and scholars are responsible for this fabrication, which hinges on two very slim pieces of evidence: 1) the fact that in making the shield of Achilles Hephaistos put Okeanos around the rim (*Iliad* 18.607), and 2) the common epithet applied to Okeanos, *apsoroos*, "back-flowing," i.e., "flowing back into itself." That is certainly not much to go on, but somehow the bizarre image of a disk-shaped earth surrounded by a flowing river of water has become firmly rooted in the minds of Homeric scholars, maybe because they only think of Greece as sun-drenched, failing to imagine Greece under a canopy of stars, much like we now imagine WWII to have taken place in black & white rather than in color. Whatever the reason for this remarkable blind spot, it persists, and has led to further groping in the dark. G. S. Kirk and J. E. Ravens, wondering where such a bizarre notion could have come from, suggest that the conception of an encircling river "originated further east, in the great river-civilizations of Egypt and Mesopotamia, and was somehow introduced into Greece and given a specific Hellenic form."[3] One wonders why these eminent scholars ignored the judgment of their predecessor at Cambridge, A. B. Cook, who stated,

> We can hardly avoid [the] conclusion [that] 'back-flowing Okeanos,' the very 'source of the gods,' was not originally a terrestrial river forming the circumference of a discoidal earth, but, as E. H. Berger has maintained, a celestial stream of stars. I should indeed venture to suppose that in pre-Greek times, before the rise of geographical speculation, the river Okeanos simply meant the Galaxy.[4]

The Milky Way, flowing in a great arc over our heads, the lower portion of the circle invisible below the horizon, is, according to the ancient lore, the river we all must cross to reach that "farther shore" where our ancestors await our coming, whether known as Okeanos, or Dante's Acheron, or the River Jordan. Gilgamesh poled across it to reach Dilmun and his ancestor Untnapishtim. Multitudes have been ferried across by the mercenary Charon, whose payment accompanied the dead into their graves. Sometimes the spirits of the dead have been able to use a bridge, the Chinvat Bridge of Iranian mythology, for example, guarded by two dogs, or the Asgard Bridge of Norse mythology before it became confused with the rainbow.[5]

It is Odysseus's fate that he must visit the Land of the Dead before he can reach his homeland. Circe gives him some rather detailed directions:

> Set up your mast, spread the white sail and sit down in the ship. The North Wind will blow her on her way; and when she has brought you across the River of Ocean, you will come to a wild coast and to Persephone's Grove, where the tall poplars grow and the willows that so quickly shed their seeds. Beach your boat there by Ocean's swirling stream and march on into Hades' Kingdom of Decay. There the River of Flaming Fire [*Puriphlegethon*] and the River of Lamentation [*Cocytus*], which is a branch of the Waters of Styx, unite round a pinnacle of rock to pour their thundering streams into Acheron.
>
> (10.505-515)

Except for some symbolic vegetation, this description fits remarkably well with what we can observe in the vicinity of Orion and the Milky Way. The several rivers are nowadays united into one, the river constellation called Eridanus, which ends far down in the southern hemisphere at a star called Achernar (from Arabic *Al Ahir al Nahr,* "the End of the River").[6] The name Eridanus first appeared in the third century B.C. *Phaenomena* of Aratos. Before this the constellation was called simply "the river," or sometimes "the river of Orion."[7] Where this name came from is uncertain. No Greek etymology is apparent, leading de Santillana and von Dechend to suggest that it may be derived from the Sumerian *Eridu*, "the confluence of the rivers."[8] The constellation is made up of a long, winding string of stars, most of them no brighter than the fourth magnitude. Because of its length and the relative dimness of its many stars the course of the river has been imagined to run several different ways. Aratos described Eridanus as it is commonly displayed on contemporary celestial maps, although even these do not always agree. Eratosthenes places the star Canopus in the constellation, thereby giving the river a different direction.[9] The celestial map provided by the authors of *Hamlet's Mill* shows a configuration of the constellation that conforms remarkably to the description given by Homer.[10] [figure 5.1] On this map we can see two rivers, one of them forked, wind through the stars and eventually unite.

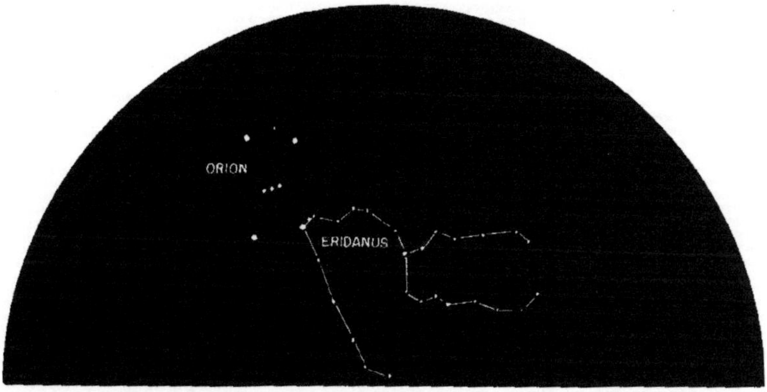

Figure 5.1
Eridanus

The pinnacle of rock mentioned by Circe seems to have been an important landmark along the path to the Land of the Dead, since it is mentioned again in Book 24 of the Odyssey, where the landmarks of our final journey are, curiously, given in reverse order:

> Past Ocean Stream, past the White Rock, past the Gates of the Sun and the region of dreams they went . . .
> (24.11,12)

The location of the White Rock has perplexed those who have tried to sort out a geography in the Odyssey.

Some have identified it as a promontory on the island of Leukas near Ithaca. The White Cliffs of Dover have also been suggested.[11] To locate the White Rock in the heavens is equally perplexing. It may have been imagined as the star Rigel located at the left knee of Orion. The two celestial rivers do come together here, as Circe insisted. The Gates of the Sun are, of course, the Gates of Heaven discussed in Chapter 3.

When it comes to Odysseus's own account of his voyage to interview the souls of the dead, neither the confluence of the rivers nor the White Rock are mentioned, and he does not cross the River of Ocean. Instead, the text reads as though Odysseus and his men remained on the near shore of Okeanos:

> [The ship] brought us to the deep-flowing River of Ocean and the frontiers of the world, where the fog-bound Cimmerians live in the City of Perpetual Mist. When the bright Sun climbs in the sky and puts the stars to flight, no ray from him can penetrate to them, nor can he see them as he drops from heaven and sinks once more to earth. For dreadful Night has spread her mantle over the heads of that unhappy folk. Here we beached our boat and . . . made our way along the banks of the River of Ocean till we reached the spot that Circe had described.
>
> (11.13-22)

Unlike the Laestrygonians, the Cimmerians were an actual people who moved from the north into Asia Minor in the eighth century B.C. The name "Cimmerian" in Homer's poem may simply have meant "dwellers in darkness," not referring to the actual tribe at all. But in order to place them on the surface of the earth, in a land where the sun is never seen and night is perpetual, some scholars have chased them into the winter latitudes of Norway. The solution to the whereabouts of the Cimmerians which is least offensive to common sense is to locate them on the shore of Okeanos where Homer says they live, at the *naksatra* called *ardra*, which means, appropriately, "moist." This *naksatra* is the star we know as Betelgeuse.[12] *Ardra* may have been named "moist" by the Indo-Europeans because this star does in fact lie on the shore of the great celestial river. Or it may have at one time been the focus of a tale involving mists and clouds, as Homer suggests. There is no way to know.

Although Odysseus is told by Circe that he is to cross Okeanos and find his way to the Halls of Hades and Persephone, he does not travel quite so far, for reasons that remain unexplained. He does not encounter the dreaded hound that guards the entrance to the Land of the Dead, Cerberus, the famous dog with three heads, otherwise known as the Dog Star, Sirius. Instead, he digs a trench near the shore of Okeanos and fills it with blood so that the spirits of the dead will approach and speak to him. We are not told where they come from, where in heaven Hades is. It is likely (if I may be allowed to speculate) that this underworld realm was once imagined to be below the horizon in that portion of the Milky Way that remains invisible to us in the northern hemisphere. The position of Sirius, or Cerberus, in the southern sky does at least suggest such a location for Hades. The myth of Hercules dragging Cerberus out of Hades could have been based on the fact that Sirius had gradually moved higher and higher in the southern sky as a result of the shift in the polar axis, the phenomenon that also causes the precession of the equinoxes. Once rising on the eastern horizon to the south of our post marking the winter solstice sunrise, Sirius rose directly over it in 3000 B.C., and now rises well to the north of it. It has long been recognized that most, and perhaps all, of Hercules's exploits contain celestial references,[13] and such a shift in the rising point of the brightest star in the heavens would not have gone unnoticed in a tradition of astronomical observation that apparently lasted for several thousand years.

The circle of the Milky Way intersects the ecliptic near Orion in the eastern quarter of the heavens, and again in the region of the constellations Scorpio and Sagittarius in the western quarter. Ithaca, the homeland of Odysseus, lies *pros zophon* (9.26) "towards the dark quarter," i.e., the west. Thus, before arriving home Odysseus will have to cross the great expanse of the night sky that lies within the circle of the Galaxy on the path known as the *devayana*, and encounter the constellations we know as Gemini, Hydra, Cancer, Leo, Corvus, and Virgo. This next leg of Odysseus's voyage will not feature the overlapping of images previously seen in our discussion of the eastern quarter, since this sequence of asterisms was never troubled by the shifting equinox. Some overlapping will appear again when we arrive at those stars that indicated the autumn equinox in ages past, but crossing the great celestial expanse of summer in the wake of Odysseus we shall find that the correspondences between the Odyssey and the Vedic *naksatras* are straightforward, however obscure they might first appear. Table 5.1 continues the list of *naksatras* and the adventures of Odysseus to which they correspond:

TABLE 5.1

Homer's term	*naksatra*	*meaning*	*asterism*
Cimmerians	ardra	moist	Betelgeuse
River of Ocean	-	-	Milky Way
Sirens	punarvasu	restore favor	α β Geminorum
Scylla	tisya	??	γ δ Θ Cancri
Planctae	aslesas	embracer	δ ε η ρ σ ς Hydrae
Thrinacie	maghas	giver of plenty	Regulus
Fig tree	phalguni	??	β δ Θ Leonis
Calypso	citra	bright	Spica
Scherie	nistya	outcast	Arcturus
Double olive	visakhe	two branches	α β Librae

CHAPTER SIX

The Northern Path

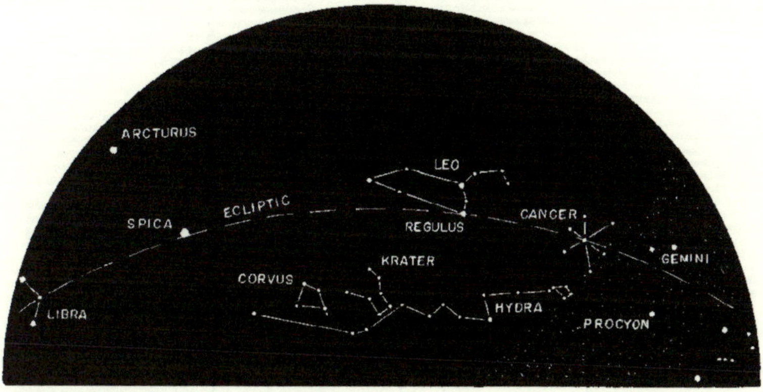

Figure 6.1
The Northern Path

The route of Odysseus along the northern path, through the sequence of constellations that rise with the dawn during the summer months, follows the course taken by the Argonauts on their return from Colchis with Medea and the Golden Fleece. After stopping at Circe's palace, the Argonauts sailed past the Sirens, Orpheus singing a counter-melody to drown out their enchanting song. They then sailed through the

Wandering Rocks rather than encounter Scylla and Charybdis, later passing Thrinacie where they could see Phaethusa and Lampetie, the daughters of Helios, looking after the sheep and cattle of the Sun, and on to the island of the Phaeacians and the hospitality of Alcinous and Arete.

The voyage of the Argonauts is a pre-Homeric saga, and was obviously familiar to Homer and his audience (12.70). Some scholars believe that this leg of Odysseus's voyage was based on legendary accounts of the Argonauts.[1] Other scholars believe that the voyage of the Argonauts and the voyage of Odysseus both stem from the same body of legends.[2] It is likely that both voyages have roots in traditional astronomical lore, each perhaps recording different celestial phenomena.

THE SIRENS

Like Jason's, Odysseus's first encounter after leaving Circe is, as she predicted, with the Sirens. Following Circe's stratagem, Odysseus stops the ears of his men with beeswax and has them tie him to the mast so that he alone might hear the enchanting song of the Sirens without succumbing to their spell. The wind dies down, the men furl the sail, and in the calm Odysseus hears the song.

Our own word "siren," meaning an enchanting woman, can be traced back to Homer. But all attempts to trace the etymology of the word beyond Homer to an earlier Greek root have failed.[3] In Homer's time the Sirens were not the voluptuous women of more recent imagination; they were great birds with the faces of women, "half human and half bird in form," according to the author of the *Argonautica* (4.891-901), and were once the companions of Persephone before she was carried off to rule in Hades. Surrounded by the skeletons of their victims, they appear in Homer as demons of the underworld,[4] luring those travelers who have made it past the entrance to Persephone's realm back to the kingdom of the dead.

If the sequence of Odysseus's adventures does correspond to the Vedic list of *naksatras* we should find the Sirens in the constellation Gemini, specifically the two bright stars now called Castor and Pollux, α and β Geminorum. The Greeks apparently did not identify the Dioscuri with this pair of stars until the fourth century B.C.,[5] the association of "twins" with this constellation having originated in Babylonia.[6] The Vedic name, *punar vasu*, meaning "restore wealth,"[7] offers no clues by which we can associate the Sirens with these two stars, and at this point in our search we come up nearly empty-handed. The only hint of a connection lies in Homer's use of the dual form when mentioning the Sirens—they are a pair, like these two stars.

This is, admittedly, a very slender piece of evidence, one which by itself hardly counts as convincing. The correspondences between the sequence of *naksatras* listed in the Vedas and the sequence of adventures recounted in the Odyssey are clearly evident in the neighborhood of the equinoxes, as we have already seen among the stars that marked the vernal equinox, and as we shall see again once we arrive at the autumn equinox. But the stars that lie along the path between them apparently did not enjoy the constant attention that those two corners of the year received, and what remains is fragmentary and circumstantial. This does not mean that there is no hope of reconstructing the celestial path of Odysseus as he sails across the broad expanse of heaven from spring to autumn. We can simply place the two sequences side by side, and with the shreds of evidence that have survived, it is possible to match the two, adventure for asterism. To do this we have to trust that the underlying structure of the

Odyssey has remained intact, and given the evidence of the fidelity of the oral tradition presented so far, the Odyssey warrants our trust.

Once past the Sirens the dilemma for Odysseus is to choose between two courses. One leads to the Wandering Rocks (Planctae), remeniscent of the Clashing Rocks (Symplegades) which the Argonauts encountered, the other leading to the dreadful Scylla, a creature with six long necks, on each neck a head with triple rows of teeth. The two *naksatras* following *punar vasu* in the Vedic list are *pusya* (Cancer) and *aslesas* (the group of stars marking the head of the constellation Hydra). These two *naksatras* are exceptionally close to one another, and both, along with punar vasu, lie along the edge of the Milky Way. If we assume that *aslesas* is the stellar home of the Wandering Rocks, then we should find Scylla in the constellation Cancer.

SCYLLA

Cancer is a constellation remarkable for the faintness of its stars and its equally obscure ancestry. It is known to us as the crab, an image that probably originated with the Egyptians. This image did not become familiar in Greece until the fifth century B.C.[8] when it took the place of an earlier Greek figure, imagined as two asses (γ and δ Cancer) standing of either side of a manger (the nebula at the center of the constellation). Aratos, Eratosthenes, and Ptolemy all called this nebula *phatnē,* "manger," the same word used in the Gospels for the birthplace of Our Lord. According to Platonist philosophy Cancer was the gate through which souls descended from heaven into human bodies.[9] The Accadian designation for this constellation was *nangaru*, meaning "carpenter" or "woodworker."[10]

The nebula at the center of Cancer is really a

cluster of faint stars (Galileo counted more than thirty through his telescope; modern astronomers have counted more than 150). This nebula was called *Nephelion* "Little Cloud" by Hipparchus, and *Akhlus* "Little Mist" by Aratos. Circe describes Scylla as protruding from a "misty cavern" which leads down to Erebus. The nebula could well have been imagined as her cavern, and since this cavern lies on the edge of the Milky Way it could easily be imagined as leading down to Erebus. The configuration of the whole constellation, a nebula surrounded by six brighter stars, does resemble Scylla's six heads on long necks protruding from her misty cavern. A coin minted at Agrigentum in Sicily between 413 and 408 B.C.[11] combines the images of Scylla and the crab of Cancer, which may have preserved the ancient identification of Scylla with this asterism. [figure 6.2]

Figure 6.2

Coin showing Scylla and the Crab of Cancer

A curious detail in Circe's description of Scylla—she has a bark like a new-born puppy's (12.86)—contains a play on words: *Skylla* (Scylla) and *skylax* (puppy). Even though the pun is obvious, it strikes one as a feature too incongruous to include among Scylla's other features, her twelve feet and six heads and triple rows of teeth, raising the suspicion that there may be a reason to include the puppy's bark beyond mere word-play. Cancer lies directly on the ecliptic, yet it is the most inconspicuous figure in the Zodiac. In order to locate the constellation when it was close to the horizon, ancient astronomers watched for brighter stars that rose simultaneously with it. In later Greek astronomy and astrology these simultaneously rising stars were called *paranatellonta*, from *para* = beside, and *anatello* = rise. The Lesser Dog Star, Procyon, whose rising coincides with Cancer, was apparently associated with the constellation in this way. Latin speakers called this star Catulus, "the puppy," and according to Robert Graves it was Scylla's emblem in heaven.[12]

THRINACIE

After the encounter with Scylla, Odysseus and his men come ashore on Thrinacie, the Island of the Sun. This island can be identified as the bright star Regulus. The name for Regulus in the Atharva Veda (14.1.12) is *magha*. This seems to be a later substitution for *agha*, found in an identical passage in the Rig Veda (10.85.13). In this Rig Veda hymn, which celebrates the marriage of Surya, daughter of the Sun, we read,

> In the sign of the angry ones (*Agha*) they kill the cattle.

Sanskrit scholars puzzle over this line, guessing that the cattle are killed for the wedding feast. Homeric scholars likewise puzzle over the killing of the cattle of Hyperion, the deed so vehemently warned against by Tiresias and Circe, the deed that delays the homecoming of Odysseus and prevents the return of the rest of his men. On one level the slaughter is seen as a violation of the laws of hospitality, a recurring theme in the Odyssey, paralleled by the slaughter of Odysseus's own livestock by the suitors back in Ithaca. On another level many commentators have suggested that the numbers in each herd, 350 cattle and 350 sheep, are a veiled reference to the number of days, 354, in a lunar year, just as the number of hogs remaining in Eumaeus's keeping, 360 (14.20), may be a reference to the solar year of 365 days. Walter Burkert suggests that myths about cattle belonging to the sun may have their roots in remotest antiquity, even beyond the emergence of the Indo-European language.[13] We may never discover the prehistoric source which might explain the slaughter of cattle in the Rig Veda or the slaughter of Hyperion's cattle in the Odyssey, but at least we can locate the Island of the Sun at Regulus with some confidence.

CHARYBDIS

Sailing away from Thrinacie all of his men are lost in a great storm, and Odysseus, clinging to the wreckage of his ship, must face the whirlpool alone. Circe has told Odysseus that "A great fig-tree with luxuriant foliage grows upon the crag, and it is below this that dread Charybdis sucks the dark waters down." (12.103-4) And sure enough, as Odysseus approaches the whirlpool clinging to the shattered remains of his ship,

> Charybdis was beginning to suck the salt water down. But as she did so, I was flung right up to the great fig-tree, on which I got a tight grip and clung like a bat. I could find no foothold to support me, nor any means of climbing into the tree, for its roots were far away below, and the great long branches that overshadowed Charybdis stretched high above my head.
>
> (12.431-436)

Why should Homer insist that it was a fig tree, rather than some other species, which grew above the whirlpool? The answer is that a fig tree once flourished in the night sky. It was located in the hind quarters of our constellation Leo, and was made up of three stars, β δ Θ Leonis.[14] This *naksatra* was called *phalguni,* a word of uncertain meaning, later divided into *purva-phalguni* and *uttarah-phalguni* (meaning former and latter).

A tale preserved by Ovid in his *Fasti* (II, 243-265) associates the constellations Corvus (the Raven) and Krater (the Goblet) with this ancient fig tree. As the story goes, Apollo, who was preparing a feast for Zeus, sent the Raven for some water. The raven picked up a golden goblet, flew off, and eventually came upon a fig tree whose fruit was not yet ripe. Because the raven had a special liking for figs, he waited for the fruit to ripen. After some time he was finally able to feast on the fruit, and once he had satisfied his appetite he picked up a water-snake and returned to Apollo. Naturally Apollo, who had been waiting all this time for a little water, was angry. The raven protested that it wasn't his fault—the snake had withheld the waters. Apollo saw through the lie, and as punishment condemned the raven ever after to go thirsty whenever figs were ripe. Ovid concludes the tale by pointing to the contiguous positions in the sky of the Raven, the Goblet, and the Water-snake (Hydra). Ovid does not mention the astronomical location of the fig tree and probably didn't know of it, since it had been superseded by the Babylonian constellation Leo long before his time. Despite this, the ancient celestial fig tree was preserved in the myth, and preserved in the Odyssey as well.

The specific detail in the Odyssey, that it is a fig tree that Odysseus clings to above the whirlpool, locates him near the stars that now make up the hindquarters of our constellation Leo. The whereabouts of the whirlpool itself, however, remains a mystery. Unless one is able to construe the constellation Corvus into the shape of a whirlpool, an implausible exercise at best, no configuration of stars below Leo would seem to fill the bill. The authors of *Hamlet's Mill*, Giorgio de Santillana and Hertha von Dechend, offer the only possible solution: "We think that the whirlpool stands for the 'ecliptical world' marked by the whirling planets."[15]

CALYPSO

In contrast to the other mythical figures Odysseus encounters, Calypso appears only in the Odyssey. The *Argonautica* makes no mention of her, Jason having sailed from the Island of the Sun directly to Scherie, the land of the Phaeacians. Her name does appear here and there in ancient texts among the lists of sea nymphs called Nereids—in the Homeric Hymn to Demeter (442), in Hesiod's catalogue of the many daughters born to Thetys and Okeanos (*Theogony*, 359), and again in the list of Nereids born to Nereus and Doris given by Apollodorus (I.2.7). But only in the Odyssey is she given both a local habitation and a name, as well as a different father, Atlas. Many things about her are mysterious—her isolation, her name, which means something like "hiding," the name of her island, Ogygia, which means something like "ancient," her remoteness from both mankind and the gods.

In Homer's description of Calypso's island one detail has baffled the geographers of the Odyssey: Ogygia is located *t' omphalos esti Thalassēs*, "where the navel of the sea is"(1.50). No such isolated island has ever been found at the center of the Mediterranean; however, this term describes perfectly the *naksatra Citra*, "bright," the star Spica in the constellation Virgo.

Spica is easy to locate in the night sky, being the only bright star in a quarter of the heavens remarkable for the sparseness of its stars. Solitary and remote, Spica lies midway on the celestial path across the expanse encircled by the Milky Way, half-way between Circe's Island of the Dawn and Ithaca, "farthest out to sea in the West" (9.25). Spica is also the closest bright star to the center of the great circle of Okeanos, the actual center lying slightly above Spica in the constellation Coma Berenices.[16] Spica is thus not only the navel of the sea, but the navel of Okeanos as well.

Odysseus is destined to spend seven years on the island of Ogygia. Finally freed from the embrace of the lonely Calypso by the intervention of Pallas Athene, Odysseus sails away, homeward bound on his makeshift raft. The beautiful, passionate Calypso is left behind, with only her weaving to keep her occupied in her solitary and ageless immortality. Odysseus, like sailors ever since, never looks back:

> It was with a happy heart that the good Odysseus spread his sail to catch the wind and used his seamanship to keep his boat straight with the steering-oar. There he sat and never closed his eyes in sleep, but kept them on the Pleiads, or watched Bootes slowly set, or the Great Bear, nicknamed the Wain, which always wheels round in the same place and looks across at Orion the Hunter with a wary eye. It was this constellation, the only one which never bathes in Ocean's Stream, that the wise goddess Calypso had told him to keep on his left hand as he made across the sea. So for seventeen days he sailed on his course, and on the eighteenth there hove into sight the shadowy mountains of the Phaeacians' country, which jutted out to meet him there. (5. 269-280)

This is the sole reference in the Odyssey to constellations we recognize today, and the only Homeric constellations recognized by a long succession of commentators, one even going so far as to insist that these were the only constellations known to Homer and his contemporaries, the others having not yet "become separated from the anonymous throng circling in the 'murk of night.'"[17] This same scholar goes on to say that "Homer's scanty astronomical notions . . . are precisely such as would arise among a people in an incipient stage of civilization, simple, direct, and child-like in their mode of regarding natural phenomena, yet incapable of founding upon them any close or connected reasoning."[18] These remarks were made before the turn of the century, and more recent scholars have been only slightly more generous in their evaluations of the intellectual sophistication of the ancient Greeks. D. R. Dicks insists that the Greeks had not distinguished the planets from the fixed stars until the fifth century B.C.,[19] and refused to entertain the possibility that the Greek Dark Age might have brought an end to a traditional body of astronomical knowledge as thoroughly as it brought an end to the Mycenaean culture itself.[20]

Another common misconception is that this passage gives clear evidence of celestial navigation in Homer's time. According to H. L. Lorimer, for example, "Odysseus is quite able to understand the sailing orders given him by Calypso in terms of constellations—single stars are not of much use for navigation"[21]—a remark which would astound anyone who has sailed the seas at any time during the past 3000 years. Calypso told Odysseus to keep the Great Bear (the Big Dipper) on his left, which is an almost meaningless piece of navigational advice—"face east" is all it could ever mean. Heading in such a vague direction might work for a very short voyage, but if Odysseus really sailed the Mediterranean for eighteen days following this advice he could have landed almost anyplace between Egypt and Corfu. Ernle Bradford, the author of *Ulysses Found,* gets around this by insisting that what Homer meant was that Odysseus should keep the Bear 45 degrees to port of the bow, not "anywhere on the left, from ahead to almost astern."[22] But even rewriting Homer solves little, because in the course of a single night the Bear would have revolved around the pole in a circle with a diameter of about 60 degrees.

This passage has also inspired at least one scholar to extrapolate the season from the periods of visibility of Bootes and the Pleiades.[23] These two constellations lie opposite one another in the sky, so that as one sets, the other rises. To see them both at the same time would have been possible for only a little more than an hour after one of the two rose, yet this would have occurred at some time during the night the year round, except for a period in the Spring when the Pleiades would have been invisible, and a like period in the Fall when the sun obscured the starlight of Bootes.

The seasonal and navigational considerations that these lines give rise to are meaningless when we realize that this same passage occurs almost word for word in the famous description of the Shield of Achilles in the Iliad (18.487-89), the only difference being that the Iliad names the Hyades instead of Bootes. Rather than being a fragment of navigational lore, this passage is merely a descriptive set-piece, and one apparently in wide circulation, as one line reappears in the *Works and Days* (615) of Hesiod.[24]

A further problem with this passage has to do with the reference to the constellation of the Bear being the only one which never bathes in Okeanos (5.275). This has commonly been understood to mean that the Bear, being a circumpolar constellation, never dipped below the horizon. But this expression, containing the word *oie*, "alone, the only one," has puzzled commentators since Aristotle because the Bear is obviously not the only circumpolar constellation. Aristotle took the expression to mean "the best known standing 'alone',"[25] while recent commentators imagine that Homer didn't know of any other circumpolar constellations.[26] But since we now know that Okeanos refers to the Milky Way, the expression is not all that puzzling. Except for the Bear, the constellations embossed on the Shield of Achilles can all be found in the Milky Way. The description is valid in the Iliad, but when the passage was transferred to the Odyssey with Bootes substituted for the Hyades the description ceased to make sense, since Bootes, like the Bear, also lies outside the Milky Way.

Rather than being a record of the earliest Greek notions of astronomy, this brief reference to the stars during Odysseus's troubled voyage from Ogygia to Scherie is obviously a late interpolation, and has more to do with the Shield of Achilles than it does with the actual astronomy that lies just below the surface of the Odyssey. This will become especially apparent when we follow Odysseus to his next destination, the *naksatra* that once marked the Autumn equinox.

CHAPTER SEVEN

✣ *Olive Trees at the Corners of the Year*

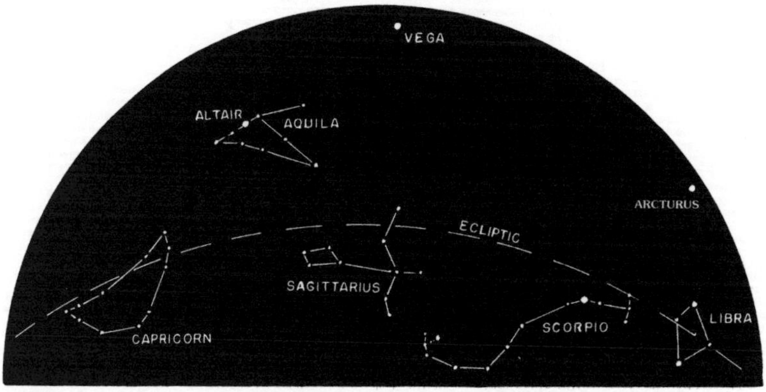

Figure 7.1
The Western Quarter

Odysseus washes up on a beach naked and near death. By now he has lost all his ships and all his men. He has nothing left except his life, and that hangs by a thread. Encrusted with salt from the sea and with a veil wrapped around his waist like an umbilical cord, he is about to be reborn. Scherie, where he has arrived, lies on the border between two worlds, one magical, one real. He is about to leave one realm and enter the other. Soon he will have everything back: his herds, his son, his wife, his father, his kingdom. He crawls beneath a double olive tree and covers himself with leaves "as a lonely crofter in the far corner of an estate buries a glowing brand under the black ashes to keep his fire alive . . ."(5.486)

This is the second of the three strange olive trees in the Odyssey. It is worth asking why Homer should include such a curious detail as an olive tree with two branches growing from the same stem, one a wild olive, the other domesticated, when any simple bush would have hidden Odysseus for the night. The answer is that Odysseus has arrived at the autumn equinox, where the *pitryana* meets the *devayana.* In the heavens this point was at one time marked by two stars, α and β in the constellation Libra. The Vedic name of this *naksatra* is *visakhe,* meaning "branched"[1] or "forked,"[2] or "possessed of two branches,"[3] an asterism that signaled the autumn equinox when the vernal equinox took place in Aries. Aries, as we saw earlier, was the celestial home of Polyphemus, whose staff of olive wood marked one corner of the year. This strange olive tree marked another, and as we shall see, the olive tree around which Odysseus built his bed marked a third, the winter solstice. The summer solstice is not designated in the Odyssey, for reasons which remain unexplained, and there seems to be no clear reason why the poets of the Odyssey chose the olive tree to indicate the equinoxes and winter solstice. Yet the fact that these three olives, all of them peculiar, occur precisely at those *naksatras* which stood at the corners of the year in the second millennium B.C. cannot be mere coincidence.

This correspondence between the olive trees in the Odyssey and the stars that once located the equinoxes and solstice depends on our matching the sequence of Odysseus's adventures with the sequence of *naksatras* given in the Vedas, adventure for *naksatra*. So far the correlation has worked out without doing violence either to the Odyssey or the Vedic list of *naksatras*. But an alert reader might now object that we have skipped one *naksatra*. The *naksatra* next in line after Spica, Calypso's Isle, is not, in fact, the asterism called "possessed of two branches." It is the bright star Arcturus, in Sanskrit *nitsya*, meaning "outcast."[4] All of the other naksatras can be found close to the ecliptic, while Arcturus lies far above the others, much closer to the pole, "ever traversing the northern way," as the *Santikalpa*[5] describes this *naksatra*.

The first thing Homer tells us about the inhabitants of Scherie, the Phaeacians, is that they have not always lived where they do now. They once lived in Hypereia, near the tribe of Cyclopes, but sometime in the past they had pulled up stakes and moved in order to get away from their quarrelsome neighbors. Once again Homer provides us with a piece of information that at first seems totally irrelevant, and we are led to wonder why it should be mentioned at all. If by now we might suspect one more stellar reference we would not be disappointed. Homer's name for the former home of the Phaeacians, Hypereia, means "the place above," a perfect description of Arcturus.

When we look at the position of Arcturus in the night sky we can only wonder that Arcturus was included among the twenty-eight *naksatras* in the first place. Because it lies so much closer to the pole than the other *naksatras*, it rises well ahead of the previous *naksatra* in the Vedic list and sets well after the *naksatra* that follows. We can be certain that for ancient astronomers heliacal risings and settings were an essential tool. Yet the position of Arcturus made it useless as a cipher on the face of their celestial timepiece. In the Odyssey Arcturus is called "late-setting Bootes"[6]

(5.272); in more ancient times and in more northerly latitudes, say 45 degrees, Arcturus would have barely set at all, only dipping below the northern horizon and rising again almost immediately. Whatever the reason for its initial election to the ranks of the *naksatras*, the evidence in the Odyssey suggests that it was eventually expelled. But even though Arcturus no longer served as a *naksatra*, its original connection with the Phaeacians was not forgotten. This connection is further exemplified in the story of Nausicaa.

While Odysseus is sleeping under the double-branched olive, Athene comes to the princess Nausicaa in a dream and tells her to procure a wagon and a pair of mules and take the dirty clothes to the river to wash them. She does as the goddess commands, and accompanied by her maids drives the wagon to the washing-pools near where Odysseus lies sleeping. As Homer has told us some 250 lines earlier (5.273), a wagon can be found in the night sky: it is the Wain, otherwise known as the Great Bear or the Big Dipper. By continuing the curve of the "handle" of the Dipper across the sky one can easily locate Arcturus. It is the only bright star in the neighborhood, and has long been associated with the constellation of the Bear. The name Arcturus comes from *Arkhtos*, "bear," and *ouros*, "guardian." The other name for Arcturus, Bootes, derives from *bous*, "ox," this star representing the man driving the oxen that pull the wagon.

Like the other celestial references in the Odyssey, the mention of the wagon strikes us as peculiar. We would expect the girls to pile the clothing on their heads and walk to the stream, but the poets of the Odyssey have placed the stream too far away for that, justifying the introduction of the celestial wagon. We would not expect that a princess in a wealthy kingdom would drive a mule-cart herself, yet this was probably a way to get around the even more ludicrous spectacle of a princess driving an ox-cart. And Nausicaa rather than a male servant had to drive the cart because tradition dictated that Odysseus be found by a group of maidens, the Pleiades. In this scene the whole sky is brought into play: the double-branched olive rising in the east, the maidens setting in the west, and between them the wagon driven by Nausicaa.

The Phaeacians entertain Odysseus for several days, during which time he recounts the adventures that took him across the northern path, from the Spring to the Autumn equinox, a tale that takes up Books 9 through 12 of the Odyssey. His generous hosts then carry Odysseus in one of their magic ships across the seas to his own island, and deposit him, sound asleep, in a very strange cave:

> When the brightest of all stars came up, the star which often ushers in the tender light of Dawn, the ship's voyage was done and she drew near to Ithaca. Now in that island is a cove named after Phorcys, the Old Man of the Sea, with two bold headlands squatting at its mouth so as to protect it from the heavy swell raised by rough weather in the open and allow large ships to ride inside without so much as tying up, once within mooring distance of the shore. At the head of the cove grows a long-leaved olive-tree and nearby is a cavern that offers welcome shade and is sacred to the Nymphs whom we call Naiads. This cave contains a number of stone basins and two-handled jars, which are used by bees as their hives; also great looms of stone where the Nymphs weave marvelous fabrics of sea-purple; and there are springs whose water never fails. It has two mouths. The one that looks north is the way down for men. The other, facing south, is meant for the gods; and as immortals come in by this way it is not used by men at all. (13.93-112)

This mysterious cave with two entrances, one for mortals and one for immortals, a cave containing, of all things, jars of honey, caught the attention of Porphyry of Tyre, a neo-platonist philosopher living in the third century A.D. He composed a treatise explaining the cave in the allegorical fashion commentators on the Odyssey have so often adopted. According to Porphyry, the cave is a symbol of the cosmos, the Naiads symbolic of souls descending into material existence, the stone kraters and amphoras symbolizing this material existence, the sea-purple cloth woven on stone looms symbolic of flesh wrapping around bones, the honey symbolic of pleasure. What Homer intended by the puzzling two entrances to the cave Porphyry understood to be the tropics of Cancer and Capricorn, the former the gate through which souls descend to their earthly existence, the latter the one by which they depart again, being, as they are, immortal.[7]

Porphyry was compelled to an allegorical interpretation of this passage because it is, on the surface, both implausible as a literal description and incomprehensible as a casual fiction. While it is hard for us to accept Porphyry's allegorical reading, we must agree that something seems to lie beneath the surface here. And what lies beneath the surface is again the ancient astronomical language, and again a description of the autumn equinox, where the two paths, the *pitryana* and the *devayana*, came together. As discussed earlier, the *pitryana* was the path of men, the *devayana* the path of the gods, or as Homer says, "the immortals."

Like the vernal equinox, the autumn equinox had shifted position in the heavens over the course of centuries, and like the vernal equinox, traces of the ancient astronomical code describing the autumn equinox can be found in more than one celestial location. We have recognized one equinox in Libra, the *naksatra* whose name means "possessed of two branches." This cave, the more ancient, describes the autumn equinox in Scorpio, more precisely the *naksatra anuradhas,* "propitious,"[8] at the time when the vernal equinox took place in the Pleiades. Even though this *naksatra* no longer served as the autumn equinox, the oral tradition preserved the complex of symbols once associated with this asterism, just as the tradition preserved several references to the spring equinox—the Island of the Dawn or the paths of day and night coming close together, for example—long after the precession had moved the equinox to another location in the heavens. Apparently a long poem in the oral tradition may be added to, embellished with more lively descriptions, but once any detail has taken root in the song it is rarely discarded or forgotten.

The cave with two mouths, one for men and one for the gods, contains several curious details. Why the Naiads wove on looms of stone defies explanation. But the stone basins and two-handled jars infested with bees are reminiscent of Indra's honey-filled footprints which marked the corners of the year, as discussed in Chapter Two. Whatever the reason for this curious symbol, it does seem to have survived in several guises. The footprint of the summer solstice may be contained in the words of Samson, "Out of the [Lion] came something sweet" (Judges 14.14). The long-leaved olive tree, another of the peculiar details provided by Homer, is, as we have already seen, one more indicator of the autumn equinox, this one eventually superseded by the olive with two branches.

It is also significant, in celestial terms, that Athene appears to Odysseus outside this cave undisguised, in her full glory. Odysseus himself, and a host of scholars after him, have wondered why Athene failed to come to the aid of her favorite earlier in his travels. The celestial explanation is that Athene is associated with the star Antares, only now making her heliacal appearance.

I realize that such an assertion is likely to be met with incredulity. Most scholars would agree with A. B. Cook's conclusion[9] that Athene, unlike Zeus or Poseidon, began her life as a local goddess, associated with the acropolis at Athens long before the Greek-speaking peoples arrived there. Hence her symbols—the olive, the snake, the owl—which are all denizens of the acropolis. She is indigenous to Greece, born of its soil. The invading Greek-speakers did not bring her name with them; at least no Greek etymology has been found for her name. However, Athene's other appellation, *Pallas*, is a Greek word, meaning a maiden, and according to W. K. C. Guthrie it is likely that a maiden-goddess came along with the invading Greeks and was later associated with the indigenous spirit of the acropolis.[10] Hence Athene's conflicting attributes: she wears the armor and carries the spear of a warrior, yet she is the goddess of household crafts, particularly of weaving and spinning. She is virgin, yet her symbol of the serpent suggests that she once was a goddess of fertility. Her belated appearance at this juncture indicates that the maiden-goddess Pallas whom the Greek-speakers brought with them had her origin in the night sky. The celestial qualities of this maiden-goddess were eventually forgotten, though the Homeric poems still preserve traces of her life in the heavens.

Antares is the bright, first-magnitude star in the constellation Scorpio. Its name, meaning "the rival of Ares," or Mars, refers to its reddish hue, similar in color to the red planet. In the Iliad Athene is also the rival of Ares. By itself the name Antares is not sufficient reason to connect Athene with this star. But Homer supplies another clue:

> Athene . . . left the pleasant land of Scherie, crossed the barren seas, and came to Marathon and the broad streets of Athens, where she entered the great palace of Erechtheus... (7.78-81)

At the end of the nineteenth century the British archaeologist Francis Penrose investigated the orientation of many Greek temples. Many conclusions he drew regarding the dates of construction of the temples have since been shown to be erroneous, but the accuracy of his work with surveying instruments remains undisputed.[11] He found that the ancient Erechtheum was aligned on the point on the horizon at which Antares could be seen to set in the eleventh century B.C.[12]

The tardy appearance of Athene, showing herself only when Odysseus arrives at the autumn equinox, suggests that the two ancient paths in the heavens, the *devayana* and *pitryana,* were once thought to be ruled by a divine power. In the Odyssey these powers are given to Poseidon and Athene. At the beginning of the *devayana,* at the vernal equinox, Odysseus defeats Poseidon's son Polyphemus, and the wrath of Poseidon follows him all the way to Scherie and the autumn equinox. Here at the beginning of the *pitryana* Athene appears, and her assistance enables Odysseus to defeat the suitors and regain his kingdom. Thus the great circle of the ecliptic is seen to have two rulers, the scepter changing hands at the equinoxes.

The cave with two mouths, one for gods and one for mortals, is one of the oldest celestial symbols in the Odyssey, and can be traced back to the Indo-European division of the ecliptic into two paths joined at the equinoxes. Curiously, after Odysseus leaves the cave and sets about re-establishing order in his kingdom, celestial references of another kind appear. These do not correspond to the sequence of *naksatras;* they seem to be later in origin and more immediately visual, yet they bear the unmistakable stamp of astronomy.

The first occurs during the slaughter of the suitors. The dead lie in heaps, the groans of the dying echo throughout the hall, and in the midst of the melee Phemius, the bard, begs for his life:

> He laid the hollow instrument on the ground half-way between the mixing-bowl and the silver-studded chair, and then ran up to Odysseus, flung his arms around his knees and poured out his plea...
>
> (22.340-3)

This detail about the lyre strikes one as curious, especially in the context of the mayhem and confusion going on all around. Rather than demonstrating "the care that a musician always shows for his instrument," as W. B. Stanford suggests,[13] this passage pinpoints the winter solstice and the festival of Apollo on the celestial clock. When we look at the night sky we will not find the lyre (the constellation Lyra) half-way between the mixing bowl (Krater) and the silver-studded chair (Cassiopeia). However, in the Greek latitudes in the second millennium B.C. during the night preceding the winter solstice, Krater rose about 90 degrees east of due north, four hours later Lyra rose 60 degrees east of due north, and in another four hours Cassiopeia rose 30 degrees east of due north. The celestial lyre was thus half-way between the mixing bowl and the silver-studded chair both in time and in the location of its rising. Furthermore, these three constellations could be seen to rise in the course of a single night only at the time of the winter solstice. This curious detail in the narrative is actually a double-check on the southern-most rising of the sun to fix the time of the solstice.

Another celestial image appears in the bizarre execution of the twelve women who had consorted with the suitors. Odysseus tells his son Telemachus to

> take the women out of the hall between the round-house and the great wall of the courtyard, and use your long swords on them, till none are left alive to remember their loves and the hours they stole in these young gallants' arms... (22.441-5)

Telemachus does take them out to the place between the round house and the courtyard wall, but instead of using his sword on them, he hangs them all at once, in a complicated (and not quite believable) fashion:

> he took a hawser which had seen service on a blue-bowed ship, made one end fast to a high column in the portico, threw the other over the round-house, and pulled it taut at such a level as would keep their feet from touching the earth. And then, like doves or long-winged thrushes caught in a net across the thicket where they come to roost, and meeting death where they had only looked for sleep, the women held their heads out in a row, and a noose was cast round each one's neck to dispatch them in the most miserable way. For a little while their feet kicked out, but not for very long. (22.465-73)

The twelve unfortunate servants can still be seen today as part of the constellation Pisces, strung on a line between the round-house (Cetus) and the wall of the courtyard (the square of Pegasus). [figure 7.2]

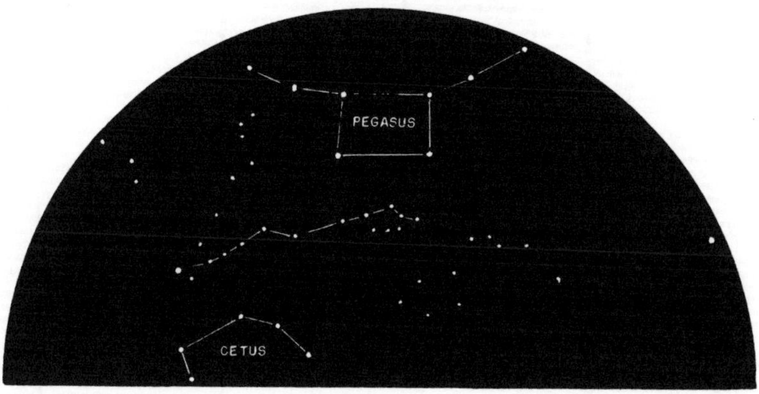

Figure 7.2
Pisces

Finally, we come to the last of the mysterious olive trees, the corner-post of Odysseus's famous bed. Homer recounts how Penelope, as adept at clever stratagems as her husband, uses a ruse to discover whether the man seated before her is really Odysseus. She suggests that the bed should be moved outside the bedroom and, as she had hoped, Odysseus bursts into a rage:

> Who, if you please, has moved my bed elsewhere? Short of a miracle, it would be hard even for a skilled workman to shift it somewhere else, and the strongest young fellow alive would have a job to budge it. For a great secret went into the making of that complicated bed; and it was my work and mine alone. Inside the court there was a long-leaved olive tree, which had grown to full height with a stem as thick as a pillar. Round this I built my room of close-set stone-work, and when that was finished, I roofed it over thoroughly, and put in a solid, neatly fitted, double door. Next I lopped all the twigs off the olive, trimmed the stem from the root up, rounded it smoothly and carefully with my adze and trued it to the line, to make my bed-post. This I drilled through where necessary, and used as a basis for the bed itself, which I worked away at till that too was done, when I finished it off with an inlay of gold, silver, and ivory, and fixed a set of purple straps across the frame. There is our secret, and I have shown you that I know it. What I don't know, madam, is whether my bedstead stands where it did, or whether someone has cut the tree-trunk through and shifted it elsewhere.
>
> (23.184-204)

Happily, the bed still stood in the place where Odysseus left it twenty years before. The olive tree had remained intact, and along with it, the celestial framework so ingeniously worked out and put to memory and transmitted over generations. For this olive tree marked the winter solstice, just as the other olive trees in this tale marked the equinoxes. If it had been displaced, it would mean that the celestial framework had come unhinged. There is more at stake in this clever ruse of Penelope than her assurance that Odysseus is who he says he is.

When the vernal equinox stood in Aries, marked by the olive staff of Polyphemus, the autumn equinox stood in Libra, marked by the double olive that Odysseus takes shelter beneath, and the winter solstice stood in the constellation Delphinus, the Dolphin. But Delphinus is a very faint constellation, having no stars brighter than the fourth magnitude. The heliacal rising of the first magnitude star Altair would have been used instead to mark the winter solstice. Altair was known in Sanskrit as *Asvattha*,[14] the name given by the Hindus to the sacred tree of life, whose roots were in heaven.[15] [figure 7.3]

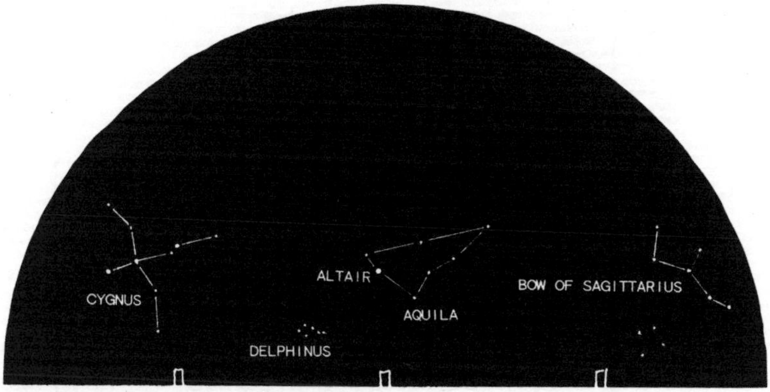

Figure 7.3

Winter Solstice Constellations

In earlier ages the winter solstice would have occurred in the constellation we know as Pegasus, which was imagined by the Vedic poets as a couch or bed.[16] When the precession shifted the solstice to

Asvattha one of the astronomer/poets must have come up with the solution of combining the ages-old bed with a living tree—surely the most curious of all the curious details in the Odyssey. This solution may have engendered the notion of placing olive trees as the other corners of the year, providing Polyphemus with an olive staff and placing olive leaves on the *naksatra* called "possessed of two branches."

The ancient astronomers, inheritors of a tradition that spoke of stars which once heralded the equinoxes and solstices but did so no longer, knew that *Asvattha*, too, would not endure forever as a herald of the winter solstice, that another age would eventually come to an end. What they did not know was when this would happen. Hence the underlying anxiety of the poem. Will Odysseus return or not? Does the bedstead stand where it did, or has someone cut the tree-trunk through and shifted it elsewhere?

Odysseus can only spend this one night with his wife before setting out again on a journey, this one decreed by Tiresias when Odysseus visited him in the Land of the Dead. He must take an oar and wander from city to city until he comes across a people who know nothing of the sea, never use salt with their food, and who would call the oar a "winnowing fan." There he must plant the oar in the earth and sacrifice a ram, a bull, and a breeding boar to Poseidon. Could it be that Odysseus once shared the destinies of the sun or moon or planets in their ceaseless wanderings among the stars?

CHAPTER EIGHT

❖ *Who was Odysseus?*

And we're lost out here in the stars.
Little stars, big stars,
blowing through the night,
And we're lost out here in the stars.
Kurt Weill

We have followed the trail of Odysseus across the night sky, beginning at the vernal equinox in Aries, travelling along the northern path of the sun and moon to the autumn equinox in Libra, and ending at the winter solstice in Aquila. Along the way a number of very curious details scattered throughout the poem have been explained as residues of an ancient system of astronomy. Three olive trees marked the corners of the year: the olive stake which puts out the eye of Polyphemus, the double olive that shelters Odysseus on Scherie, the olive around which Odysseus constructed his bed. Circe's island of the dawn is another name for the asterism that once marked the vernal equinox. The paths of day and night coming close together at

Telepylus are the *devayana* and *pitryana* of Indo-European astronomy, which come together again at the cave with two entrances, one for mortals and one for gods. The island of goats near the cave of the Cyclops, the incestuous household of Aeolus, the slaying of the giant stag, the name of the magic plant "mōly," the fig tree above the whirlpool, Nausicaa and her maidens–all of these are residues from the language of ancient astronomy. Okeanos is now seen as the great circle of the Milky Way, with Calypso's isle, "the navel of the sea," as its center. The curious reference to Phemius placing his lyre between the silver-studded chair and the mixing bowl describes a phenomenon seen only at the winter solstice. The unfaithful servant women all hanging by their necks in a line can still be seen in the night sky.

None of these curious and seemingly gratuitous details has been explained before, yet taken together we find that they all contribute to a larger pattern, one apparently contrived by astronomers living several thousand years before Homer. The adventures of Odysseus described in books 9-12 cover one-half of the circle of the ecliptic, from the vernal equinox to the autumn equinox. The homecoming of Odysseus covers the other half of the ecliptic, from the autumn equinox to the winter solstice and beyond to the constellation Pisces, bringing us back to Aries and the vernal equinox again. The evidence that a succession of asterisms took turns marking the equinoxes indicates that some version of an astronomical poem began as early as 3000 B.C., with the vernal equinox in Taurus. The configuration of the heavens found in the Odyssey, placing the vernal equinox at the *naksatra bharanis* in Aries, existed between 2200 and 1200 B.C.

We have an answer to our original question, *What is the Odyssey about?* It is "about" the stars. But this answer only raises further questions: Was the

Odyssey constructed from remnants of an ancient astronomical poem which somehow got mixed up with legends about a hero of the Trojan War? Or did Odysseus figure in the original poem, one of the celestial wanderers—sun, moon, or planets—that the ancients sought to keep track of? If the former, our task is come to an end, and perhaps we ought to leave it at that. However, having come this far, the possibility that Odysseus once lived in the night sky warrants a few paragraphs, speculative at best, and admittedly inconclusive.

The time of Odysseus's return is given repeatedly and precisely: he arrives in Ithaca "in the twentieth year" (2.175; 17.327; 23.102; 23.170; 24.322) "while one moon is waning and another is on the rise" (14.162; 19.307) and kills the suitors as they celebrate the festival of Apollo, which was probably a solstice festival. These details, among others, suggest one of the wandering heavenly bodies. But which one?

Neither the sun nor the moon fit the narrative pattern of the Odyssey, since neither one could be said to return after an absence of twenty years. Of the five visible planets only Saturn is a likely candidate. Saturn completes his journey around the ecliptic in 29.458 years. Obviously if Saturn rose heliacally at the winter solstice, he would not be there twenty years later. However, if the heliacal rising of Saturn at some time coincided with the vernal equinox, the encounter with Polyphemus, in the twentieth year another heliacal rising of Saturn would coincide with the winter solstice, the marital bed. If one were to object that the twenty-year span pertains to Odysseus's absence from Ithaca, there is another possible scenario. If at the winter solstice Saturn were to appear at the bow of Sagittarius, probably the great bow that only Odysseus could string, twenty years before he could have been seen near the constellation Pisces, "between the round-house and the great wall of the courtyard." Both of these are in Ithaca, which, in the astronomical design, represents the southern path, stretching across the heavens from the autumn equinox to the vernal equinox.

There are other reasons for entertaining the possibility that Odysseus once played the role of Saturn. Since twenty-eight *naksatras* were identified along the ecliptic, Saturn comes close to visiting one *naksatra* each year. And for nearly eight months of each year retrograde motion would cause Saturn to appear to linger at a particular *naksatra* before resuming his progress to the next one.

At the beginning of the voyage Odysseus's men kill 108 goats, and at the end of the journey Odysseus kills 108 suitors. It is a curious number, all the more curious when we learn that 108 is a sacred number in India. There are 108 classical Upanishads, 108 bhakti prayers, 10800 stanzas in the Rig Veda, 10800 bricks in the Hindu fire altar. The Saivite rosary contains 108 beads, the beautiful gopis who danced with Lord Krsna numbered 108, there are 108 shrines of special sanctity in Hinduism. Even today women of the Golog tribe in Tibet wear their hair in 108 braids.[1] The number appears in the lore of Greece as well. Heraclitus is said to have believed that the Great Year, marked by all the planets returning to the positions they held at the beginning, comprised 10,800 years.[2]

We are all familiar with the sacred numbers three, or seven, or twelve, but why 108? Various puzzled commentators have tried to work out an origin for this number. One solution held that it was the product of the twenty-seven *naksatras* multiplied by the four phases of the moon;[3] another took the five planets plus the sun and moon, added the two phases of the moon for a total of nine, and then multiplied by the twelve signs of the zodiac.[4] A more plausible answer is that this number was associated with the planet Saturn. Saturn makes his journey around the ecliptic in 10,759.2 days according to modern calculations. Using the more symmetrical number 10,800, or 30 years of 360 days, would result in an error of only a little more than one degree in the position of Saturn after thirty years' time.

In Vedic lore the number 432 occurs as frequently as the number 108. Each of the 10,800 stanzas of the Rig Veda contains forty syllables, 432,000 all told. The present epoch, the Iron Age, known as the *Kali Yuga*, will last 432,000 years. In the description of the end of the world in the Icelandic Poetic Edda, Valhalla has 540 doors, and each door will release 800 warriors for the final battle, 432,000 in all. This number, too, may have its origin in ancient astronomy, since one cycle of Jupiter takes 4,332.6 days by modern calculation–in round numbers 4,320 days, or twelve years of 360 days.

The sacred number 108 and its companion 432, numbers which appear throughout the Indo-European world, imply a year of 360 days. Taken together these numbers were apparently used to construct a cosmology. In this system, one cycle of Saturn took thirty years, one cycle of Jupiter took twelve years, and, perhaps relevant to our inquiry, conjunctions of Saturn and Jupiter took place, like clockwork, every twenty years, the length of time since the roaming Telemachus last saw his father.

❖ ❖ ❖

Like so many mysteries in the Odyssey, we may never know if Odysseus did in fact begin his long career as the planet Saturn, but in a way it does not matter. He now wanders in our thoughts and in our dreams, no longer luminous, but clothed in our human condition. Although Odysseus wears, like the scar on his thigh, the remnants of this ancient astronomy, what Homer wants us to see is a man, one both cursed and blessed by the gods, and a glorious and magical world, now gone from the face of the earth. The poem, begun in an attempt to probe the mysteries of Time—Time made visible by the silent, circling stars overhead—culminated in the portrait of a world, like our own, haunted by the passage of time.

Achilles is dead, and Patroclus, and Hector, and Agamemnon—the age of heroes is over. The winged words soar into a sunset, and like a sunset the Odyssey is suffused with sadness. The sadness of time passing, of old comrades dead, the sadness of absence and of longing, for a father, a husband, a home. Yet it is a world redeemed by the splendor of the telling, and redeemed by a story that exists outside of time: imprisoned in the cave until the stone is rolled away on the third day, he will return from the land of the dead, a beggar in the palace, to punish the wicked and restore justice to the land. We will know him by his scar.

It is well to keep in mind that the Odyssey is a work of incomparable beauty, not merely a puzzle to be solved. Giorgio de Santillana, while acknowledging the poem's astronomical roots, had such reverence for the artistry of the Odyssey that he cautioned against the very kind of investigation I have here undertaken. "It is not fitting," he wrote, "to try to reduce to a 'scheme' what remains a prodigious and subtle work of art, the limpidity and immediacy of which should not be spoiled. It is unfortunately a common prejudice that to work out high theoretical allusions contained in the text reduces the text to an irrelevant conundrum ..."[5]

Throughout the writing of this book I have been haunted by those words. I think de Santillana knew that the astronomical structure of the Odyssey could be explained if one would only spend some time studying the stars. He was also aware that the astronomy of the Odyssey is only the bones, the skeleton. The blood is in the flesh, in the music sung in an ancient tongue. Even though this new perception of the almost unimaginable antiquity of the Odyssey makes the epic still more marvelous than we had ever suspected, there is a risk in dwelling on the astronomical dimension to the exclusion of its rich human drama. Homer and his predecessors have turned the poem into something more than astronomy, something even more beautiful than the constellations shining in the night sky.

APPENDIX

The Celestial Location of Ortygia

The etymology of Sirius, like that of Orion, is uncertain. Homer does not use this name for the star, calling it instead "Orion's dog" in the Iliad (22.26-29). The term Sirius (Seirios) first appears in Hesiod's *Works and Days* (609-11) and is used both for this brightest of all the stars and for the sun itself. These two designations, Sirius and Orion's dog, seem to have been peculiar to the Greeks. The Egyptians knew the star as Sothis. It was of primary importance to them, as its heliacal rising heralded the annual flooding of the Nile, and also signaled the beginning of their solar year. The zodiac taken from the ceiling of the Egyptian temple at Dendera, which dates from Roman times, depicts Sirius as a cow riding in a boat with a star between her horns. The goddess Satit stands nearby, aiming a bow and arrow at this star. The Chinese constellation constructed from these same stars is remarkably similar to the Egyptian, the bow and arrow formed from stars in our present Canis Major and Argo, and aimed at Sirius, the celestial jackal, or wolf, T'ien-lang.[1] The Babylonians, too, saw the same bow in the night sky, but here Sirius was the arrow, KAK.SI.DI, rather than the target.[2] [figure 9.1]

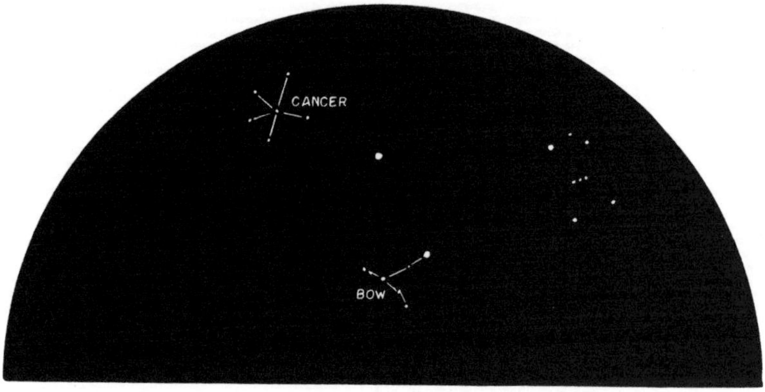

figure 9.1
The Celestial Bow with Sirius

It does not seem far-fetched to locate Artemis and her bow here at Sirius, along with the other slayer of Orion, Rudra. By doing so, the argument over the location of Ortygia can be laid to rest. Ortygia is the birthplace of Artemis and her brother Apollo, and also the place where she killed Orion. Ortygia literally means "Quail Island," and in antiquity this was thought to refer to the island of Delos, because quail were common there, or perhaps the island districts of Syracuse.[3] But neither of these locations corresponds to the description given by Eumaeus, the swineherd, in Book 15 of the Odyssey. Eumaeus begins the sad story of his life with, "There is an island called Syrie—you may have heard the name—over against Ortygia, where the sun turns in his course."(15.403-4) There is, obviously, no tropic, or turning-place, of the sun at either Delos or Syracuse; however, the sun most definitely did turn in his course at the constellation Cancer, which rose simultaneously with Sirius and marked the summer solstice towards the end of the second millennium B.C.—hence our term "Tropic of Cancer." And Eumaeus's home, Syrie, is, to me at least, closer to Hesiod's Seiros than it is to Syracuse.

If we assume that the bow of Artemis is to be found in the constellation Canis Major and Ortygia is to be found at the constellation Cancer, another celestial coincidence will also appear: Sirius in Canis Major and Vega in the constellation Lyra were, in the late second millennium B.C., seen on the horizon at the same time, whether rising or setting. When Sirius rose, Vega set; when Vega rose, Sirius set—an appropriate coincidence for the lyre of Apollo and the bow of Artemis, twins at birth.

NOTES TO THE INTRODUCTION

1 Gilbert Murray's description of this period in Greek history has become a classic in its own right. See *The Rise of the Greek Epic*, 4th ed. (1934; reprint, New York: Oxford University Press, 1960), 27-60.

2 Rhys Carpenter, *Discontinuity in Greek Civilization* (Cambridge: Cambridge University Press, 1966). His theory is disputed by H. Wright, Jr., *Climate Changes in Mycenaean Greece, Antiquity,* xlii, 166(1968), 123; affirmed by climatologists R.A. Bryson, H.H. Lamb and David L. Donley, *Drought and the Decline of Mycenae, Antiquity*, xlvii, 189(1974), 46; the conclusions of the latter are disputed by O. Dickenson in *Antiquity*, xlviii, 191(1974), 228; see also John Chadwick, *The Mycenaean World* (Cambridge University Press, 1976), 192.

3 M.I. Finley, *Early Greece* (New York: W.W. Norton, 1981), 57.

NOTES TO CHAPTER ONE ✦ *The Puzzle*

1 For a more complete discussion of the allegorical interpretations of the Odyssey see Howard Clarke, *Homer's Readers,* (Newark: University of Delaware Press, 1981), 60-105; also see Robert Lamberton, *Homer the Theologian* (Berkeley: University of California Press, 1986).

2 C. Scott Littleton, *The New Comparative Mythology*, 3rd. ed. (Berkeley: University of California Press, 1982), 28-31; J. P. Mallory, *In Search of the Indo-Europeans* (London: Thames & Hudson, 1989), 182-185.

[3] Littleton, 30. Others fix the date around 1000 B.C. See P. Bosch-Gimpera, *Les Indo-Européens* (Paris: Payot, 1980), 226.

[4] F. Max Müller, *Comparative Mythology* (1856; reprint, Salem NH: Arno Press, 1977), 82.

[5] F. Max Müller, *Introduction to the Science of Religion* (London: Longmans, Green, 1873), 65.

[6] J. Menrad, *Der Urmythus der Odyssee* (Munich: J. Lindauersche Buchhandlung, 1910), 42.

[7] R. F. Littledale, "The Oxford Solar Myth," printed by A. Smythe Palmer in his 1909 edition of Max Müller's *Comparative Mythology*.

[8] Herman Fränkel, *Early Greek Poetry and Philosophy,* trans. Moses Hadas and James Willis (Oxford: Blackwell, 1975), 48.

[9] Joseph Campbell, *The Masks of God: Occidental Mythology* (New York: Viking, 1964), 157-177.

[10] Samuel Butler, *The Authoress of the Odyssey* (1897; reprint, Chicago: University of Chicago Press, 1967).
[11]Victor Bérard, *Navigations d'Ulysse* 4 vols.(Paris: Colin, 1927-29).

[12] Ernle Bradford, *Ulysses Found* (New York: Harcourt, Brace & World, 1963).

[13] Mauricio Obregon, *Ulysses Airborne* (New York: Harper & Row, 1974).

[14] Gilbert Pillot, *The Secret Code of the Odyssey* (New York: Abelard-Schumann, 1972).

[15] A. Rousseau-Liessens, *Géographie de l'Odysée* (Brussels: Brepols, 1961).

[16] Tim Severin, *The Ulysses Voyage* (New York: G. P. Dutton, 1987).

[17] Howard Clarke, *Homer's Readers* (Newark: University of Delaware Press, 1981), 251.

[18] Walter Burkert, *Structure and History in Greek Mythology and Ritual* (Berkeley: University of California Press, 1979), 33. Thirty-six of these folktales are summarized by Sir James Frazer in the appendix to his translation of Apollodorus in the Loeb Classical Library (Cambridge: Harvard University Press, 1921).

[19] Rhys Carpenter, *Folktale, Fiction and Saga in the Homeric Epics* (Berkeley: University of California Press, 1951), 18.

[20] W. B. Stanford, ed., *The Odyssey of Homer* 2d ed., Vol. 1 (New York: St Martins Press, 1971), notes to XXIII 190*ff.*, 399-400.

[21] cf. Walter J. Ong, *Orality and Literacy* (London: Methuen, 1982), 140; Erik A. Havelock, *Preface to Plato* (Cambridge: Harvard University Press, 1963), 291-292.

[22] For a further discussion of this notion of 'code' see Giorgio de Santillana, *The Origins of Scientific Thought* (Chicago: University of Chicago Press, 1961), 7-20.

[23] Gerald S. Hawkins, *Stonehenge Decoded* (Garden City: Doubleday, 1965).

[24] R. J. C. Atkinson, "Megalithic Astronomy--A Prehistorian's Comment," *Journal for the History of Astronomy*, vol. 6, pt. 1, p. 51.

NOTES TO CHAPTER TWO ☆ *The Celestial Framework*

[1] Richmond Lattimore, trans., *Hesiod* (Ann Arbor: University of Michigan Press, 1959).

[2] Richard Hinckley Allen, *Star names: Their Lore and Meaning* (1899; reprint, New York: Dover Publications, 1963), 256.

[3] Arthur Anthony Macdonell and Arthur Berriedale Keith, *Vedic Index of Names and Subjects* (1912; reprint, Delhi: Motilal Banarsidass, 1967), 409.

[4] Kathaka Samhita xi.3; xviii.14; Taittiriya Samhita ii.3,5,1-3; iii.4,7,1; Vajasaneyi Samhita xviii.40.

[5] Wendy Doniger O'Flaherty, ed. and trans., *The Rig Veda* (New York: Penguin Books, 1981), 226.

[6] O'Flaherty, 213.

[7] Bal Gangadhar Tilak, *The Orion: or Research into the Antiquity of the Vedas* (Poona City: Tilak Bros., 1893), 12.

[8] O'Flaherty, 43-44.

[9] R. C. Zaehner, ed. and trans., *Hindu Scriptures* (London: J. M. Dent, 1966), 100.

NOTES TO CHAPTER 3 ✯ *The Gates of Heaven*

[1] Anton Scherer, *Gestirnnamen bei den Indogermanischen Völkern* (Heidelberg: Carl Winter, 1953), 162, 166, 188; André Le Boeuffle, *Les Noms latins d'Astres et de Constellations* (Paris: Société d'Édition "Les Belles Lettres," 1977), 180.

[2] All passages quoted from the Odyssey are from the translation by E. V. Rieu (Penguin Books, 1946).

[3] Robert Graves, *The Greek Myths* (Penguin Books, 1960), 41d.

[4] Aratus, *Phaenomena,* 634 ff.

[5] Hyginus, *Poetic Astronomy*, ii. 34.

[6] Graves, 41a,b.

[7] The root of *rudra* is rud, meaning "red." Many ancient authors, including Ptolemy, state that Sirius is red in color. Apparently Sirius changed from red as late as classical times to its present white, puzzling contemporary astronomers. See Kenneth Brecher, "Sirius Enigmas," *Technology Review,* 80(2): 52-63 (December 1977).

[8] Scherer, 147.

[9] Richard Hinckley Allen, *Star Names: Their Lore and Meaning* (1899; reprint, New York: Dover Publications, 1963), 389.

[10] The latest period during which any asterism can announce the equinox by its heliacal rising depends on its brightness, the configuration of the horizon, and the clarity of the atmosphere. Aldebaran stood at the equinoctial point in 2870 B.C. The last visible heliacal rising would have been several hundred years earlier. The Pleiades stood at the equinoctial point in 1830 B.C., and 41 Aries in 900 B.C. Their final dates would have been around 2200 B.C. and 1200 B.C. respectively.

[11] As translated by W. B. Stanford in his notes to 10.82ff. in *The Odyssey of Homer* 2d ed. vol. 1 (New York: St. Martin's Press, 1971).

[12] Denys Page, *Folktales in Homer's Odyssey* (Cambridge: Harvard University Press, 1973), 36ff.

[13] Bal Gangadhar Tilak, *The Orion: or Research into the Antiquity of the Vedas* (Poona City: Tilak Bros., 1893), 24.

[14] Richmond Lattimore, trans., *Hesiod* (Ann Arbor: University of Michigan Press, 1968), 168.

[15] Diodorus Siculus iii.60 and iv.27; Graves 133.3.

16 Giorgio de Santillana and Hertha von Dechend, *Hamlet's Mill* (Boston: Gambit, 1969), 166.
17 For a more complete discussion, see de Santillana and von Dechend, 90.

18 de Santillana and von Dechend, 232-233.

19 de Santillana and von Dechend, 137 ff..

NOTES TO CHAPTER FOUR
⁂ *Further Remnants of the Precession*

1 (1.2.15) quoted in Howard Clarke, *Homer's Readers* (Newark: University of Delaware Press, 1981), 251.

2 See Mary Barnard, *The Mythmakers* (Athens, OH: Ohio University Press, 1966), Chapter 12 "The Dancing Stars."

3 Hesiod, *Works and Days,* 383.

4 Ovid, *Fasti,* iv. 170ff.

5 Wendy Doniger O'Flaherty, ed. and trans., *The Rig Veda* (New York: Penguin Books, 1981), 77 & 82.

6 Clarke, 89.

7 André Le Boeuffle, *Les Noms latins d'astres et de constellations* (Paris: Société d'Édition "Les Belles Lettres," 1977), 121.

8 Le Boeuffle, 121; Anton Scherer, *Gestirnnamen bei den indogermanischen Völkern* (Heidelberg: Carl Winter, 1953), 123.

9 Lattimore translation.

[10] Scherer, 141-143.

[11] Scherer, 159; Arthur Anthony Macdonell and Arthur Berriedale Keith, *Vedic Index of Names and Subjects* (1912; reprint, Delhi: Motilal Banarsidass, 1967), 419.

[12] Bartel L. van der Waerden, *Science Awakening II* (New York: Oxford University Press, 1974), 74.

[13] Le Boeuffle, 153.

[14] Scherer, 164.

[15] Scherer, 124, 162, 184; Le Boeuffle, 106-110.

[16] Denys Page, *The Homeric Odyssey* (Oxford: Clarendon Press, 1955), 11.

[17] Bal Gangadhar Tilak, *The Orion: or Research into the Antiquity of the Vedas* (Poona City: Tilak Bros., 1893), 57; Richard Hinckley Allen, *Star Names: Their Lore and Meaning* (1899; reprint, New York: Dover, 1963), 275.

[18] Evidence that it was still thriving in the centuries immediately preceding the Trojan War is contained in one of Gilbert Murray's elliptical remarks: "the ram-adventure of Odysseus in Odyssey I is depicted on an early Cretan gem and must have been current about 1500 B. C." *The Rise of the Greek Epic*, 4th ed. (1934; reprint, New York: Oxford University Press, 1960), 232.

NOTES TO CHAPTER FIVE
✧ *Okeanos and other Celestial Rivers*

[1] Herodotus, IV, 8.

[2] Aristotle, *Meteorologica* i, 9.

[3] G. S. Kirk and J. E. Ravens, *The Presocratic Philosophers* (Cambridge: Cambridge University Press, 1963), 12.

[4] A. B. Cook, *Zeus*, vol. 2 (Cambridge: Cambridge University Press, 1925), 481-482.

[5] Richard Hinckley Allen, *Star Names: Their Lore and Meaning* (1899; reprint, New York: Dover Publications, 1963), 478.

[6] Allen, 217.

[7] André Le Boeuffle, *Les Noms latins d'astres et de constellations* (Paris: Société d'Édition "Les Belles Lettres," 1977), 139.

[8] Giorgio de Santillana and Hertha von Dechend, *Hamlet's Mill* (Boston: Gambit, 1969), 257.

[9] de Santillana and von Dechend, 257.

[10] de Santillana and von Dechend, map facing p. 142.

[11] W. B. Stanford, ed., *The Odyssey of Homer* (New York: St. Martin's Press, 1971), notes to 24.11.

[12] The pronunciation of Betelgeuse ("beetle juice") is as comic to English speakers as its derivation, from Arabic *Ibt al Jauzah,* "The Armpit of the Central One."

[13] See, for example, Charles Dupuis, *Abrégé de l'Origine de tous les Cultes* (Paris: L. Tenre, 1821), ch. 5, 105-122.

NOTES TO CHAPTER 6 ✪ *The Northern Path*

[1] G. S. Kirk, *The Songs of Homer* (Cambridge: Cambridge University Press, 1962), 235.

[2] Karl Meuli, *Odyssee und Argonautica* (Basel, 1921), 25.

[3] Karoly Marot, *Die Anfang der Griechischen Literatur: Vorfragen* (Budapest: Verlag der Ungarischen Akademie der Wissenschaften, 1960), 142-149.

[4] Denys Page, *Folktales in Homer's Odyssey* (Cambridge: Harvard University Press, 1973), 86.

[5] Anton Scherer, *Gestirnnamen bei den indogermanischen Völkern* (Heidelberg: Carl Winter, 1953), 122.

[6] André Le Boeuffle, *Les Noms latins d'astres et de constellations* (Paris: Société d'Édition "Les Belles Lettres," 1977), 159.

[7] Arthur Anthony Macdonell and Arthur Berriedale Keith, *Vedic Index of Names and Subjects* (1912; reprint, Delhi: Motilal Banarsidass, 1967), 416.

[8] Le Boeuffle, 161.

[9] Robert Lamberton, *Homer the Theologian* (Berkeley: University of California Press, 1986), 70; Allen, 107.

[10] Le Boeuffle, 160.

[11] Coin number 2043 in S. W. Grose, *Catalogue of the McClean Collection of Greek Coins,* Vol. 1 (1923-29; reprint, Chicago: Obol International, 1979). Scylla also appears with the Crab on another coin, McClean catalogue number 2056, from the same time and place.

[12] Robert Graves, *The Greek Myths* (Penguin Books, 1960), 168.1.

[13] Walter Burkert, *Structure and History in Greek Mythology and Ritual* (Berkeley: University of California Press, 1979), 94.

[14] Richard Hinckley Allen, *Star Names: Their Lore and Meaning* (1899; reprint, New York: Dover Publications, 1963), 254.

[15] Giorgio de Santillana and Hertha von Dechend, *Hamlet's Mill* (Boston:Gambit, 1969), 239.

[16] Allen, 172.

[17] Agnes M. Clerke, *Familiar Studies in Homer* (London: Longmans, Green, 1892), 41.

[18] Clerke, 52.

[19] D. R. Dicks, *Early Greek Astronomy to Aristotle* (Ithaca: Cornell University Press, 1970), 167. For a rebuttal to Dicks' conclusion see Peter J. Bicknell, "Early Greek Knowledge of the Planets," *Eranos* LXVII (1970), 47-54.

[20] Dicks, 161-162.

[21] H. L. Lorimer, "Stars and Constellations in Homer and Hesiod," *Annual of the British School of Athens* 46 (1951): 92.

[22] Ernle Bradford, *Ulysses Found* (New York: Harcourt, Brace & World, 1963), 197.

[23] John A. Scott, "Assumed Contradictions in the Seasons of the Odyssey," *Classical Philology* 11(April 1916): 148-155.

[24] Another example of such a set-piece is the description of the goddess of Death, found in the Iliad (18.530ff.), repeated word for word in the Shield of Heracles (156-59) of Hesiod.

[25] Aristotle, *Poetics* (25 1461 a 25).

[26] "But more likely it reflects the comparative ignorance of astronomy in Homer's time and means 'alone' of all the

constellations recognized by Homer." W. B. Stanford, ed., *The Odyssey of Homer* (New York: St. Martin's Press, 1971), notes to 5.275.

NOTES TO CHAPTER SEVEN
✣ *Olive Trees at the Corners of the Year*

1 Wendy Doniger O'Flaherty, ed. and trans., *Hindu Myths* (Penguin Books, 1975), 357.

2 Anton Scherer, *Gestirnnamen bei den indogermanischen Völkern* (Heidelberg: Carl Winter, 1953), 157.

3 Bal Gangadhar Tilak, *The Orion: or Research into the Antiquity of the Vedas* (Poona City: Tilak Bros., 1893), 57.

4 Arthur Anthony Macdonell and Arthur Berriedale Keith, *Vedic Index of Names and Subjects* 2 vols. (1912; reprint, Delhi: Motilal Banarsidass, 1967), 417.

5 Quoted in Macdonell and Keith, 417.

6 It is likely that originally Bootes referred only to the bright star Arcturus. Later the name was used to designate the large constellation with Arcturus as its brightest star. Hesiod knew the star as Arcturus, Homer as Bootes. See Richard Hinckley Allen, *Star Names: Their Lore and Meaning* (1899; reprint, New York: Dover Publications, 1963), 92.

7 The most recent translation of Porphyry's essay is by Robert Lamberton, *Porphyry on the Cave of the Nymphs* (Barrytown, New York: Station Hill Press, 1983).

8 Macdonell and Keith, 417.

9 A. B. Cook, *Zeus* vol. 3 (Cambridge: Cambridge University Press, 1925), 224.

[10] W. K. C. Guthrie, *The Greeks and their Gods* (Boston: Beacon Press, 1955), 108.

[11] See William Bell Dinsmoor, "Archaeology and Astronomy," *Proceedings of the American Philosophical Society* Vol. 90, No. 1 (January 1939): 95-143.

[12] Penrose's figures have been reproduced in J. Norman Lockyer, *The Dawn of Astronomy* (1894; reprint, Cambridge: M.I.T. Press, 1964), 419-424.

[13] W. B. Stanford, ed., *The Odyssey of Homer* (New York: St. Martin's Press, 1971), notes to 22.340-1.

[14] Allen, 59.

[15] Margaret and James Stutley, *Harper's Dictionary of Hinduism* (New York: Harper & Row, 1977), 27.

[16] Allen, 325.

NOTES TO CHAPTER 8 ❖ *Who was Odysseus?*

[1] Galen Rowell, "Nomads of China's West," *National Geographic*, Feb. 1982, 259.

[2] David Ulansey, *The Origins of the Mithraic Mysteries* (New York: Oxford University Press, 1989), 74.

[3] Margaret and James Stutley, *Harper's Dictionary of Hinduism* (New York: Harper & Row, 1977), 130.

[4] Benjamin Walker, *The Hindu World* (New York: Frederick A. Praeger, 1968), 2:137.

[5] Giorgio de Santillana and Hertha von Dechend, *Hamlet's Mill* (Boston: Gambit, 1969), 117.

NOTES TO APPENDIX
The Celestial Location of Ortygia

[1] Giorgio de Santillana and Hertha von Dechend, *Hamlet's Mill* (Boston: Gambit, 1969), 216-217; Richard Hinkley Allen, *Star Names: Their Lore and Meaning* (1899; reprint, New York: Dover, 1963), 125.

[2] B. L. Van der Waerden, *Science Awakening II: The Birth of Astronomy* (New York: Oxford University Press, 1974), 74.

[3] W.B. Stanford, ed., *The Odyssey of Homer,* 2nd ed. (New York: St. Martin's Press, 1965-71), notes to 15.403-4 and 5.123.

WORKS CITED

Allen, Richard Hinckley. *Star Names: Their Lore and Meaning*. 1899. Reprint. New York: Dover Publications, 1963.

Apollodorus. *The Library*. Translated by Sir James George Frazer. 2 vols. Cambridge: Harvard University Press, 1921.

Aratos. *Phaenomena*. Translated by Wm I. Barnholth. Akron, OH: n.p., 1958.

Atkinson, R. J. C. "Megalithic Astronomy—A Prehistorian's Comment." *Journal for the History of Astronomy* 6, pt. 1.

Barnard, Mary. *The Mythmakers*. Athens, OH: Ohio University Press, 1966.

Bérard, Victor. *Navigations d'Ulysse*. 4 vols. Paris: Colin, 1927-29.

Bicknell, Peter J. "Early Greek Knowledge of the Planets." *Eranos 67* (1970): 47-54.

Bosch-Gimpera, P. *Les Indo-Européens: problèmes archéologiques.* Translated from Spanish by R. Lantier. Paris: Payot, 1961.

Bradford, Ernle. *Ulysses Found*. New York: Harcourt, Brace & World, 1963.

Brecher, Kenneth. "Sirius Enigmas," *Technology Review*. 80(2):52-63 (December 1977).

Burkert, Walter. *Structure and History in Greek Mythology and Ritual.* Berkeley: University of California Press, 1979.

Butler, Samuel. *The Authoress of the Odyssey*. 1897. Reprint. Chicago: University of Chicago Press, 1967.

Campbell, Joseph. *The Masks of God: Occidental Mythology*. New York: Viking, 1964.

Carpenter, Rhys. *Discontinuity in Greek Civilization.* Cambridge: Cambridge University Press, 1966.

———. *Folktale, Fiction and Saga in the Homeric Epics.* Berkeley: University of California Press, 1951.

Chadwick, John. *The Mycenaean World.* Cambridge: Cambridge University Press, 1976.

Clarke, Howard. *Homer's Readers.* Newark: University of Delaware Press, 1981.

Clerke, Agnes M. *Familiar Studies in Homer*. London: Longmans, Green, 1892.

Cook, Arthur Bernard. *Zeus: a Study in Ancient Religion*. 3 vols. Cambridge: Cambridge University Press, 1914-40.

de Santillana, Giorgio. *The Origins of Scientific Thought.* Chicago: University of Chicago Press, 1961.

de Santillana, Georgio, and Hertha von Dechend. *Hamlet's Mill: an essay on myth and the frame of time.* Boston: Gambit, 1969.

Dicks, D. R. *Early Greek Astronomy to Aristotle.* Ithaca: Cornell University Press, 1970.

Dinsmoor, William Bell. "Archaeology and Astronomy." *Proceedings of the American Philosophical Society* 90 (January 1939): 95-143.

Dupuis, Charles. *Abrégé de l'origine de tous les cultes.* Paris: L. Tenre, 1821.

Finley, M. I. *Early Greece*. New York: W. W. Norton, 1981.

Fränkel, Herman. *Early Greek Poetry and Philosophy.* Translated by Moses Hadas and James Willis. Oxford: Blackwell, 1975.

Gleadow, Rupert. *The Origin of the Zodiac.* London: Jonathan Cape, 1968.

Grant, Mary. ed. and trans. *The Myths of Hyginus*. Lawrence: University of Kansas Publications: Humanistic Studies No. 34, 1960.

Graves, Robert. *The Greek Myths*. 2 vols. Penguin Books, 1960.

Grose, S. W. *Catalogue of the McClean Collection of Greek Coins.* 1923-29. Reprint. Chicago: Obol International, 1979.

Guthrie, W. K. C. *The Greeks and their Gods* Boston Beacon Press, 1955

Havelock, Eric A. *Preface to Plato*. Cambridge: Harvard University Press, 1963

Hawkins, Gerald S. *Stonehenge Decoded*. Garden City: Doubleday, 1965.

Herodotus. *The History*. Translated by David Grene. Chicago: University of Chicago Press, 1987.

Homer. *The Odyssey*. Translated by E. V. Rieu. Penguin Books, 1946.

Kirk, G. S. *The Songs of Homer*. Cambridge: Cambridge University Press, 1962.

Kirk, G. S. and J. E. Ravens. *The Pre-Socratic Philosophers.* Cambridge: Cambridge University Press, 1963.

Lamberton, Robert. *Homer the Theologian*. Berkeley: University of California Press, 1986.

———. *Porphyry on the Cave of the Nymphs*. Barrytown, New York: Station Hill Press, 1983.

Lattimore, Richmond, trans. *Hesiod.* Ann Arbor: University of Michigan Press, 1968.

Le Boeuffle, André. *Les Noms latins d'astres et de constellations.* Paris: Société d'Édition "Les Belles Lettres," 1977.

Littleton, C. Scott. *The New Comparative Mythology* 3rd ed. Berkeley: University of California Press, 1982.

Lockyer, J. Norman. *The Dawn of Astronomy*. 1894. Reprint. Cambridge: M.I.T. Press, 1964.

Lorimer, H. L. "Stars and Constellations in Homer and Hesiod." *Annual of the British School of Athens* 46 (1951).

Macdonell, Arthur A., and Arthur Berriedale Keith. *Vedic Index of Names and Subjects*. 2 vols. 1912. Reprint. Delhi: Motilal Bannarsidass, 1967.

Mallory, J.P. *In Search of the Indo-Europeans*. London: Thames & Hudson, 1989.

Marot, Karoly. *Die Anfang der Griechischen Literatur: Vorfragen*. Budapest: Verlag der Ungarischen Akademie der Wissenschaften, 1960.

Menrad, J. *Der Urmythus der Odysee*. Munich: J. Lindauersche Buchhandlung, 1910.

Meuli, Karl. *Odyssee und Argonautica*. Basel, 1921.

Müller, Friedrich Max. *Comparative Mythology*. 1856. Reprint. Salem, NH: Arno Press, 1977.

———. *Introduction to the Science of Religion*.London: Longmans, Green, 1873.

Murray, Gilbert. *The Rise of the Greek Epic*. 4th ed. 1934. Reprint. New York: Oxford University Press, 1960.

Obregon, Mauricio. *Ulysses Airborne*. New York: Harper & Row, 1974.

O'Flaherty, Wendy Doniger. ed. and trans. *Hindu Myths.* Penguin Books, 1975.

———. ed. and trans. *The Rig Veda*. Penguin Books, 1981.

Ong, Walter J. *Orality and Literacy*. London: Methuen, 1982.

Ovid. *The Fasti, Tristia, Pontic Epistles, Ibis, and Halieuticon*. Translated by Henry T. Riley. London: George Bell & Sons, 1899.

Page, Denys. *The Homeric Odyssey*. Oxford: Clarendon Press, 1955.

———. *Folktales in Homer's Odyssey*. Cambridge: Harvard University Press, 1973.

Piggot, Stuart. *Ancient Europe*. Chicago: Aldine, 1965.

Pillot, Gilbert. *The Secret Code of the Odyssey*. New York: Abelard-Schumann, 1972.

Rousseau-Liessens, A. *Géographie de l'Odyssée*. Brussels: Brepols, 1961.

Rowell, Galen. "Nomads of China's West." *National Geographic,* Feb. 1982.

Scherer, Anton. *Gestirnnamen bei den indogermanischen Völkern*. Heidelberg: Carl Winter, 1953.

Scott, John A. "Assumed Contradictions in the Seasons of the Odyssey." *Classical Philology* 11 (April 1916): 148-155.

Severin, Tim. *The Ulysses Voyage*. New York: G. P. Dutton, 1987.

Stanford, W. B., ed. *The Odyssey of Homer.* 2d ed. 2 vols. New York: St Martin's Press, 1965-71.

Stutley, Margaret, and James Stutley. *Harper's Dictionary of Hinduism.* New York: Harper & Row, 1977.

Tilak, Bal Gangadhar. *The Orion: or Research into the Antiquity of the Vedas.* Poona City: Tilak Bros., 1893.

Ulansey, David. *The Origins of the Mithraic Mysteries*. New York: Oxford University Press, 1989.

van der Waerden, Bartel L. *Science Awakening II*. New York: Oxford University Press, 1974.

Walker, Benjamin. *The Hindu World*. New York: Frederick A. Praeger, 1968.

Zaehner, R. C. ed. and trans. *Hindu Scriptures*. London: J. M. Dent, 1966.

THE 28 NAKSATRAS

	NAKSATRA	ASTERISM	HOMER'S TERM
1	Krttikas	Pleiades	Aeolia
2	Rohini	Aldebaran	Laestrygonia
3	Mrgasiras	$\lambda\ \varphi^1\ \varphi^2$ Orionis	Circe
4	Ardra	Betelgeuse	Cimmerians
5	Purnarvasu	αβGeminorum	Sirens
6	Tisya	γδθCancri	Scylla
7	Aslesas	δεηροζHydrae	Planctae
8	Maghas	Regulus	Thrinacie
9	Purva Phalguni	δθLeonis	Fig Tree
10	Uttara Phalguni	βLeonis	Fig Tree
11	Hasta	αβγδεCorvi	
12	Citra	Spica	Calypso
13	Nistya	Arcturus	Scherie
14	Visakhe	αβLibrae	Double Olive
15	Anuradha	βδπρScorpii	
16	Jyestha	Antares	Athene
17	Mula	λυScorpii	Mōly
18	Purva Ashadha	δεSagittarii	The Great Bow?
19	Uttara Ashadha	ζσSagittarii	
20	Abhijit	Vega	
21	Asvattha	Altair	Bed Post
22	Sravistha	αβγδDelphini	
23	Satabhisaj	λAquarii	
24	Purva Bhadrapada	αβPegasi	
25	Uttara Bhadrapada	γPegasi + α Andromedae	
26	Revati	ζPiscium	
27	Asva-yujau	βγArietis	
28	Bharanis	35,39,41 Arietis	Cyclops

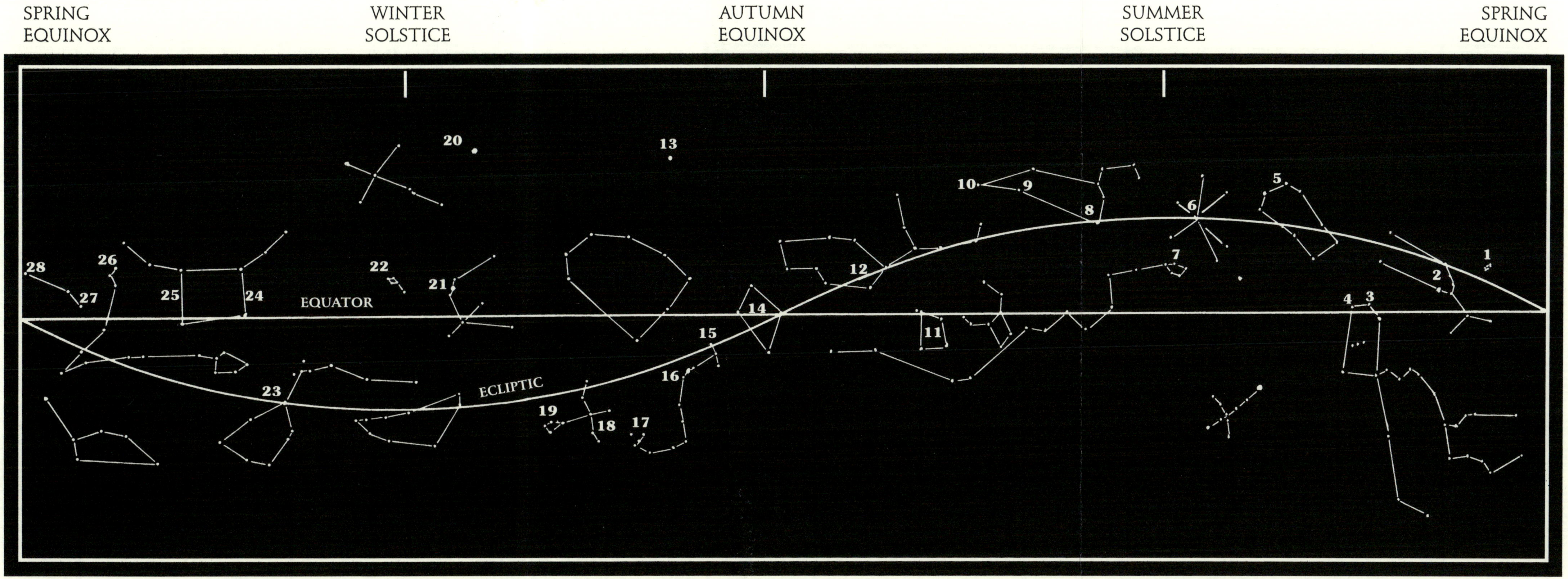

THE 28 NAKSATRAS - CONFIGURATION OF THE HEAVENS IN 1200 BC